W9-BNC-859

DISCARDED

DATE DUE

ugust 1999
, January 2000
October 2002
, with updates, January 2003
arch 2004
uly 2005

hological Association
NE
C 20002

ordered from
artment

C 20090-2984

rope, Africa, and the Middle East, copies may be ordered from
hological Association
reet
n, London
gland

be Minion by Kachergis Book Design, Pittsboro, NC

nd cover): Victor Graphics, Baltimore, MD
t and cover): Kachergis Book Design, Pittsboro, NC
duction Editors: Anne Woodworth and Jennifer Zale

ngress Cataloging-in-Publication Data
id A. M.
g your findings : a practical guide for creating tables / Adelheid
ol and Penny M. Pexman.
.   cm.
des bibliographical references and index.
1-55798-593-6 (acid-free paper)
tistics—Charts, diagrams, etc.   I. Pexman, Penny M.   II. Title.
1999
—dc21   99-24966
                                                CIP

ry Cataloguing-in-Publication Data
d is available from the British Library.

he United States of America

Presenting
Your Findings

Pres
Your

A Practical Gu

Adelheid A. M. N

Brooks –
She
Commu

American Psychological

First printing, A
Second printing
Third printing,
Fourth printing
Fifth printing, N
Sixth printing,

Published by
American Psyc
750 First Street
Washington, D

Copies may be
APA Order De
P.O. Box 92984
Washington, D

In the U.K., Eu
American Psyc
3 Henrietta St
Covent Garde
WC2E 8LU En

Typeset in Ad

Printer (text
Designer (tex
Technical/Pr

Library of Co
Nicol, Adelh
Presenti
A. M. Ni

Incl
ISBN
1. St
HA31.N5
001.4'22

British Libr
A CIP Reco

Printed in t

Contents

Acknowledgments

We would like to thank several individuals for their assistance in the creation of this book. Dr. Tony Vernon and Dr. Robert Gardner provided helpful comments regarding the entire book, and David Stanley provided excellent feedback regarding the meta-analysis chapter. The anonymous reviewers also provided excellent suggestions. The staff at APA Books were excellent and made this experience thoroughly enjoyable.

Adelheid A. M. Nicol thanks Yves Mayrand, Hilde Kunze, Dolard Nicol, and Tracy Morgan for their endless encouragement. Penny M. Pexman would like to thank Dave Pexman, Fiona and Mike Goodchild, and Sue and Warwick Pexman for their considerable support and unwavering enthusiasm for this project.

Presenting Your Findings

CHAPTER 1

Introduction

We decided to write this book when we were both graduate students at the University of Western Ontario, London, Ontario, Canada. Adelheid was trying to summarize findings for a particular analysis. She clearly needed to create a table to do so, but she was struggling to figure out what the format of the table should be. She thought there must be a guidebook or some other source that would summarize table format and was surprised to find that no such book existed. Like other students, she had to search many periodicals and statistics textbooks before locating adequate models to guide her work. One evening, she was complaining over the telephone to a friend about how she wasted too many hours looking for the proper way to make tables (when she could be doing other fun stuff like photocopying articles or running analyses). She mentioned that a reference guide definitely would be useful to have. Her friend told her that she should write the book herself (she was possibly tired of listening to her complain and was hoping that this comment would make her stop!). Adelheid thought that this was a great idea. Unfortunately, her friend wanted no part in such a task. Frustrated, she went to tell Penny about her troubles (of course, Penny first had to listen to the entire drama that led to the idea). To Adelheid's delight, Penny was keenly interested in embarking on the project (even though we were both uncertain at the time what exactly we were getting into). We decided we could solve this problem. The happy ending to this story is the book you are reading.

In writing this book, we have crafted the tool that we missed, and we hope it will be beneficial to all readers who, like us, have struggled to express statistical data correctly and elegantly. By presenting multiple examples of tables for the results of a wide range of statistical analyses, we have structured this book to make table formatting easier.

How We Created the Tables

We constructed the tables in this book by following the general guidelines for tables found in the fifth edition of the *Publication Manual of the American Psychological Association* (APA)[1] and refining the models by replicating the format of tables presented in published journal articles. For each type of statistical analysis, we looked at a broad sample of journal articles reporting that analysis. We sampled from journals reporting many different areas of research, including animal learning, clinical psychology, cognition, developmental psychology, educational research, industrial/organizational psychology, neuroscience, psychiatry, and social psychology.

The sources of reference for the tables presented here are APA's and other reputable journals. From these journals we took consensus about the common table formats for each type of statistical analysis. For some analyses, we present only one table format for results, whereas for other analyses, we present a range of formats that are commonly used.

Anatomy of a Table

Although many examples of proper table format are presented in the chapters that follow, a basic understanding of table components is important. If you find yourself grappling with the results of a particularly complicated study, knowing the functional difference between the parts such as a spanner and a boxhead will help you in developing an acceptable modification. Table 1.1 illustrates the main parts of a table,[2] which are defined as follows:

1. American Psychological Association. (2001). *Publication Manual of the American Psychological Association* (5th ed.). Washington, DC: Author.
2. Reprinted from *Publication Manual of the American Psychological Association* (5th

Table 1.1 *Sample Table Identifying Main Parts*

Table X

Mean Numbers of Correct Responses by Children With and Without Pretraining

	Girls			Boys		
Grade	With	Without	Differ-ence	With	Without	Differ-ence
Verbal tests						
3	280	240	40	281	232	49
4	297	251	46	290	264	26
5	301	260	41	306	221	85
n^a	18	19		19	20	
Mathematical tests						
3	201	189	12	210	199	11
4	214	194	20	236	210	26
5	221	216[b]	5	239	213	26
n^a	20	17		19	18	

Callouts: stubhead, column spanner, decked heads, column heads, table spanner, stub, cell, table body, notes to table.

Note. Maximum score = 320.

[a]Numbers of children out of 20 in each group who completed all tests. [b]One girl in this group gave only two correct responses.

- *Title.* Description of the contents of the table. The title should not be a word-for-word repetition of column or row heads but rather should concisely express key groups and manipulations (see Exhibit 1.1).[3]
- *Stubhead.* Heading for rows in a table.
- *Column head.* Heading for a column (entries read vertically). A column head must apply consistently to all data within that column.
- *Column spanner.* Heading for two or more columns.
- *Decked heads.* A column spanner and the column heads under it are referred to together as decked heads.
- *Table body.* The actual results (e.g., means, percentages, or *F* ratios) presented in the table.
- *Table spanner.* Subheading used for further division within the body of the table. A spanner is placed below the boxheads, centered across the entire body of the table. Spanners are used to indicate variations in data that cannot be expressed in column heads or stubheads.
- *Notes.* Used to express information needed to put the data in context. This includes spelled-out abbreviations, sample sizes and other background information about groups, or asterisks indicating significance levels.

When to Use a Table

Tables allow complex data to be expressed in a tidy format. By putting research results in a table two goals are achieved. First, details of the study are presented so they can be subjected to further analysis. Second, by removing long strings of data from the text, the study can be approached from a broad perspective, using the text to analyze trends and explore the implications of the results.

Tables should not be used when results can easily be expressed in text. If a table is unusually short (only a few columns or rows), it may be best to discuss the results within the text. Likewise, tables should be limited to the expression of data that are directly relevant to the hypotheses in the research. Detailed results that are not di-

3. From *Publication Manual of the American Psychological Association* (5th ed., p. 156) by the American Psychological Association, 2001, Washington, DC: Author. Copyright 2001 by the American Psychological Association.

rectly relevant to the hypotheses may be included as a table in an appendix.

Whether using a table in this book as a model or generating an original format for an analysis not covered here, an important rule to remember is that a good table should always stand alone. That is, the reader should not have to refer to the text for basic information needed to understand the table. Simply looking at a table should be enough for the reader to grasp the data being presented. This means that abbreviations and acronyms must be spelled out or explained and other basic information provided either in the table itself or in a table note.

A Few Caveats

It is very important to note that this book is not intended to be a statistics textbook. The purpose of this book is to provide sample tables, not advice on statistics; recommendations on how to conduct analyses are not provided. Each chapter contains only a brief description of the statistical analysis. These descriptions are intended to help the reader identify the analysis, not provide a thorough explanation of the statistical procedure. The goal was to provide a simple, visual presentation of tables along with straightforward research examples.

In some research it can be beneficial to provide figures to illustrate results. However, few figures have been included in this book; figures are presented only in the chapters on cluster analysis (chap. 6), discriminant function analysis (chap. 8), and model testing (chap. 19). Tables are the primary means of presenting research results; therefore, other modes of presenting results are not included in this book.

The examples in this book were created for the purpose of illustrating how a table can be presented. The research study examples and the data for those examples are fictional; any resemblance to actual studies is purely coincidental. In addition, all of the measures identified in this book are fictional.

Note also that readers should not consider the design of the study examples to be ideal. There may be better ways of investigating the research questions presented in the study examples. Finally, statistics presented in the tables should not be considered the *only* way those data could have been analyzed. In many cases, the data from the study examples could have been analyzed differently.

Organization of the Book

Each chapter in this book is devoted to a particular statistic. The statistics selected were included because we believe them to be the most commonly used analyses. By selecting the statistics in this way, we hoped to make the book useful for as many students and researchers as possible.

The chapters are presented alphabetically by the name of the statistic they describe. We hope that this will make it easy for the reader to find quickly a table model for whatever statistical analysis he or she is presenting. If you are using this book to understand statistical presentation in tables, we recommend that you begin by looking at the chapters on presenting means (chap. 13) and frequency and demographics (chap. 10), as the tables presenting these analyses form the basis for many other presentations.

To make this book easy to use, the chapters have a consistent format. There are five parts to each chapter:

1. a description of the statistical analysis,
2. an overview of the types of tables that are frequently presented for that particular analysis,
3. the "Play It Safe" table or tables identifying the most comprehensive presentation of the analysis,
4. one or more example studies that provide the context for the sample tables, and
5. the sample tables.

Below are descriptions of each of the five parts:

Title of Analysis

What Is It?

A brief (one- to three-sentence) description of the particular analysis is provided.

What Tables Are Used?

In one or two paragraphs the most commonly used table (or tables) for the results of the particular analysis is described.

"Play It Safe" Table

In many cases there are several alternatives presented for the format of the tables. Thus, in each chapter the particular "Play It Safe" table format is indicated. This "safe" choice is comprehensive and thus would be appropriate if the writer wanted to be as thorough as possible and was not concerned with brevity.

Example

Each chapter contains at least one study example. In some chapters, several examples are presented. These examples are intended to be straightforward and therefore do not include especially complex variables or research methodology. The intent was to make the tables easy to understand by describing an example study and using fictional data. A researcher may need to extrapolate from the sample tables if his or her statistics are more complicated.

Accompanying the description of the example study is an exhibit clearly identifying the independent and dependent variables. This is to help the reader understand and recognize the key elements of the study.

After the description of the fictional example, there are one or more tables presenting the results of the statistical analysis. Each table adheres to the guidelines outlined in the fifth edition of the *Publication Manual of the American Psychological Association*. Text notes, enclosed in oval boxes, are located within, next to, or below some sample tables. These text notes are included to point out simple modifications that could be made to a table or identify important aspects of a particular table.

We hope this book simplifies the task of creating tables for research results. Our goal was to help researchers spend more time doing research and generating ideas and less time mulling over the format of their tables.

CHAPTER 2

Analysis of Covariance

What Is It?

The analysis of covariance (ANCOVA) is an extension of the analysis of variance (ANOVA) and is used when the effects of a covariate, or uncontrolled source of variation, need to be removed from the ANOVA. An ANCOVA is used when there is one dependent variable.

What Tables Are Used?

There are two tables that are particularly relevant when presenting data analyzed with an ANCOVA: (a) a table of means and standard deviations for the dependent variable (often posttest scores) and the covariate (often pretest scores) as a function of the independent variable or variables (Table 2.1) and (b) the ANCOVA summary table (Table 2.2 or 2.3). The table of means and standard deviations and the ANCOVA summary table sometimes are combined (Table 2.4). It should be noted as well that for ANCOVA results researchers have the option of presenting either adjusted or unadjusted means, and examples of both are presented (unadjusted means in Table 2.1, adjusted means in Table 2.4).

"Play It Safe" Table

The "Play It Safe" choice for ANCOVA tables is Table 2.1 (means and standard deviations) with Table 2.2 (ANCOVA summary table).

Example 1

The example used here is that of an educational study. In this study, the researcher has developed a new aid for teaching seventh-grade students about electric circuits. The aid is a transparent circuit board that the students can experiment with on their own. The researcher wants to know whether the students learn more using the aid if they (a) explore it individually without any instruction, (b) are given written instructions about it, or (c) watch a demonstration of how it works. The researcher also wants to know whether the students learn more if an eighth-grade student tutor is assigned to help them. The researcher wants to control for the differences among the students in terms of the amount they knew about electric circuits before they began the session with the circuit board. This is measured with a presession written test (pretest). After the session with the circuit board, the students' learning will be assessed with a written test (posttest). Thus, the independent variables are instruction condition and whether a tutor assisted them during the session. The covariate is the students' presession (pretest) knowledge. The dependent variable is the students' postsession (posttest) knowledge.

Exhibit 2.1

Independent Variables
1. Instruction condition (no instruction, written instruction, or demonstration)
2. Tutor help (presence vs. absence of eighth-grade student tutor)
3. Pretest score (covariate)

Dependent Variable
1. Posttest score

Table 2.1

Table X

Pre- and Posttest Mean Scores and Standard Deviations as a

Function of Instruction Condition and Tutor Help

See chapter 13 for other examples of format for tables of means and standard deviations.

Source	Pretest		Posttest	
	M	SD	M	SD
No instruction				
Tutor help	56.12	12.11	76.90	11.22
No tutor help	54.33	11.93	73.96	12.34
Written instruction				
Tutor help	58.34	12.05	83.66	12.36
No tutor help	55.09	12.17	74.01	11.78
Demonstration				
Tutor help	64.05	11.89	86.14	10.80
No tutor help	65.12	12.34	76.44	11.24

Table 2.2

This table, along with Table 2.1 for means and standard deviations, is the "Play It Safe" table for ANCOVA tables.

Table X

Analysis of Covariance of Posttest Knowledge Scores as a Function of

Instruction Condition and Tutor Help, With Pretest Knowledge Scores

as Covariate

Source	df	SS	MS	F	ω^2
Covariate	1	39.31	9.31	4.22**	.05
Instruction condition (IC)	2	38.78	19.39	2.50*	.03
Tutor help (TH)	1	30.26	30.26	3.90**	.04
IC × TH	2	76.04	38.02	4.90**	.06
Error	54	419.04	7.76		
Total	60	573.43	9.56		

*p < .05. **p < .01.

Table 2.3

Table X

Analysis of Covariance for Instruction Condition and Tutor Help

Source	df	MS	F	ω^2
Pretest knowledge (covariate)	1	9.31	4.22**	.05
Instruction condition (IC)	2	19.39	2.50*	.03
Tutor help (TH)	1	30.26	3.90**	.04
IC × TH	2	38.02	4.90**	.06
Error	54	7.76		

*p < .05. **p < .01.

Example 2

The researchers conduct a second experiment to determine how different types of cognitive skills change as a result of exposure to the circuit board. In this experiment, a new group of 60 students participate in the circuit board session. Half of the students receive tutor help and half do not. Before the session with the circuit board, the students are tested for five cognitive skills: general problem solving, science problem solving, electronics problem solving, creativity, and spatial rotation (pretest). They are tested for these skills again following the session (posttest). The researchers perform an ANCOVA to determine how each of these skills is influenced by participation in the session while controlling for presession skill levels. The independent variable is tutor help, the covariate is presession cognitive skills (pretest), and the dependent variable is postsession cognitive skills (posttest).

Exhibit 2.2

Independent Variables
1. Tutor help (presence vs. absence of eighth-grade student tutor)
2. Presession (pretest) cognitive skills (covariate; general problem solving, science problem solving, electronics problem solving, creativity, and spatial rotation)

Dependent Variable
1. Postsession (posttest) cognitive skills (general problem solving, science problem solving, electronics problem solving, creativity, and spatial rotation)

> A full ANCOVA summary table often is not necessary, as illustrated in the following table, which is a combination of a table of means and standard deviations and an ANCOVA summary table.

Table 2.4

Table X

Adjusted Means, Standard Deviations, and Analysis of Covariance (ANCOVA) Results for Five Cognitive Skills

| Cognitive skill | Tutor help | | | | No tutor help | | | | ANCOVA | |
| | Pretest | | Posttest | | Pretest | | Posttest | | | |
	M	SD	M	SD	M	SD	M	SD	$F(1, 50)$	d
GPS	37.81	8.99	42.47	7.50	38.33	8.91	34.44	9.23	11.10***	.28
SPS	30.44	5.20	31.39	4.82	28.11	6.66	29.10	8.77	1.45	.09
EPS	20.14	6.36	28.22	9.41	18.21	7.00	22.13	7.32	7.81**	.20
Creativity	12.91	5.00	19.99	5.55	11.21	4.87	13.44	3.47	17.61***	.39
Spatial rotation	23.14	4.26	22.14	3.48	18.11	4.12	17.93	3.98	1.09	.05

Note. GPS = general problem solving; SPS = science problem solving; EPS = electronics problem solving.

**$p < .01$. ***$p < .001$.

> This table illustrates the format for a column of effect sizes.

CHAPTER 3

Analysis of Variance

One-Way Analysis of Variance: What Is It?

The one-way analysis of variance (ANOVA) is used when there is one independent variable and one dependent variable. It is used to assess the differences between two or more group means. Here the completely randomized design will be illustrated where each case or participant is represented in only one cell.

What Tables Are Used?

When there is only one analysis to report, no table is required. The means, standard deviations, and ANOVA results are presented in the text rather than in a table. One-way ANOVA results often are presented in tables for theses and for journals if there are several independent ANOVAs to be reported. The following two examples illustrate what is commonly presented in one-way ANOVA tables. For Example 1, the results of a single ANOVA are illustrated in Table 3.1. For Example 2, the results of multiple separate ANOVAs are presented (Tables 3.2, 3.3, 3.4, 3.5, and 3.6).

"Play It Safe" Table

The most comprehensive ANOVA table includes the degrees of freedom, sums of squares, mean squares, and F ratios (see Table 3.1 for a single ANOVA and Tables 3.3 and 3.4 for several independent ANOVAs). In addition, means and standard deviations are presented in the text for the results of a single ANOVA or in a table if several independent ANOVAs are presented (see Table 3.2).

Example 1

A company wishes to see the effects of three types of training programs on employees' job performance 3 months after the programs have been completed. There are three levels of the independent variable (i.e., three different training programs):

1. *Coworker program.* Employee is taught for 3 days by an experienced coworker and is provided with an information manual.
2. *Consultant program.* Employee is taught for 3 days by an external consultant and is provided with an information manual.
3. *Self-program.* Employee is provided with an information manual and learns on his or her own for 3 days.

The dependent variable is job performance. This was measured using a 7-item scale completed by each employee's supervisor. There were 40 employees in each of the three training programs.

To summarize, the study consists of one independent variable (training program) and one dependent variable (job performance).

The results of an ANOVA conducted on the data from the study described in this example are presented in Table 3.1. The means and standard deviations for the three training programs could be provided in the text; a table is not necessary to display these simple results.

Exhibit 3.1

Independent Variable
1. Training program (coworker, consultant, or self)

Dependent Variable
1. Job performance

Table 3.1

This is the "Play It Safe" table for a single ANOVA.

```
Table X

One-Way Analysis of Variance Summary for Training
Program
```

Effect size could be presented in a column to the right of the *F* ratio.

Source	df	SS	MS	F
Between groups	2	26.90	13.45	22.05**
Within group	37	22.51	0.61	
Total	39	49.41		

**p < .01.

If there is more than one one-way ANOVA to be reported, one table often is used to present the descriptive statistics (e.g., means and standard deviations), and a separate table is used to present the ANOVA results. However, some authors choose to combine the descriptive statistics with the ANOVA results.

Example 2

In this example, a company wishes to see the effects of three types of training programs on numerous attitudes and behaviors 6 months after the programs have been completed. Again, there are three levels of the one independent variable: coworker program, consultant program, and self-program. There were 300 employees in each of the three training programs.

There are seven dependent variables (all continuous variables): (a) job performance, (b) organizational commitment, (c) job commitment, (d) job satisfaction, (e) turnover intention, (f) job stress, and (g) role ambiguity. Because there are seven dependent variables, seven separate analyses must be conducted.

Exhibit 3.2

Independent Variable
1. Training program
   (coworker, consultant, or self)

Dependent Variables
1. Job performance
2. Organizational commitment
3. Job commitment
4. Job satisfaction
5. Turnover intention
6. Job stress
7. Role ambiguity

For this example, the descriptive statistics could be presented in a table such as Table 3.2. (Note that other examples of format for tables of means and standard deviations can be found in chap. 13.)

The results of the ANOVAs for the seven separate analyses could be presented in a number of ways. Tables 3.3 and 3.4 present, for all of the analyses, the degrees of freedom, sums of squares, and mean squares for both between groups and within groups. Tables 3.5 and 3.6 are examples of how descriptive statistics and ANOVA F ratios can be combined in the same table. Also presented are F ratios and significance levels (p values).

See chapter 13 for other examples of format for tables of means and standard deviations.

Table 3.2

Table X

Means and Standard Deviations for Three Training Programs and Seven Dependent Variables

Variable	Coworker		Consultant		Self	
	M	SD	M	SD	M	SD
Job performance	12.34	2.89	11.78	3.45	10.90	2.98
Organizational commitment	9.54	1.51	9.67	1.47	9.89	1.32
Job commitment	3.35	0.89	3.41	0.96	3.33	0.82
Job satisfaction	5.67	1.01	4.79	0.99	3.45	1.10
Turnover intention	1.44	0.56	1.89	0.67	2.02	0.59
Job stress	15.87	3.56	15.32	3.24	17.04	3.18
Role ambiguity	4.45	1.32	4.39	4.04	1.25	1.35

> This table, along with a table of means and standard deviations, is a "Play It Safe" table for several one-way ANOVAs.

Table 3.3

Table X

One-Way Analyses of Variance for Effects of Training Programs on

Seven Dependent Variables

> Effect sizes could be presented in a column to the right of the F ratios.

Variable and source	df	SS	MS	F
Job performance				
Between groups	2	76.04	38.02	35.87***
Within groups	297	314.82	1.06	
Organizational commitment				
Between groups	2	8.60	4.30	4.43*
Within groups	297	288.09	0.97	
Job commitment				
Between groups	2	6.84	3.42	1.67
Within groups	297	608.85	2.05	
Job satisfaction				
Between groups	2	15.32	7.66	6.78**
Within groups	297	335.61	1.13	
Turnover intention				
Between groups	2	18.56	9.28	7.42***
Within groups	297	371.25	1.25	
Job stress				
Between groups	2	42.88	21.44	9.01***
Within groups	297	706.86	2.38	
Role ambiguity				
Between groups	2	16.44	8.22	3.21*
Within groups	297	760.32	2.56	

*p < .05. **p < .01. ***p < .001.

> Inclusion of degrees of freedom in this manner is useful only if they are not the same for each analysis. Because the degrees of freedom do not vary in this table, they may be presented in parentheses following the F heading (see Table 3.4). Or, they may be presented as lettered footnotes to the table (see Table 3.12).

This table, along with a table of means and standard deviations, is a "Play It Safe" table for several one-way ANOVAs.

Table 3.4

Table X

Effects of Training Programs on Job Performance, Organizational Commitment, Job Commitment, Job Satisfaction, Turnover Intention, Job Stress, and Role Ambiguity

Variable and source	SS	MS	$F(2, 297)$
Job performance			
Between groups	76.04	38.02	35.87***
Within groups	314.82	1.06	
Organizational commitment			
Between groups	8.60	4.30	4.43*
Within groups	288.09	0.97	
Job commitment			
Between groups	6.84	3.42	1.67
Within groups	608.85	2.05	
Job satisfaction			
Between groups	15.32	7.66	6.78**
Within groups	335.61	1.13	
Turnover intention			
Between groups	18.56	9.28	7.42***
Within groups	371.25	1.25	
Job stress			
Between groups	42.88	21.44	9.01***
Within groups	706.86	2.38	
Role ambiguity			
Between groups	16.44	8.22	3.21*
Within groups	760.32	2.56	

*$p < .05$. **$p < .01$. ***$p < .001$.

Effect sizes could be presented in a column to the right of the F ratios.

In this table, descriptive statistics are presented with the ANOVA *F* ratios and effect sizes.

Table 3.5

Table X

Means, Standard Deviations, and One-Way Analyses of Variance (ANOVA) for Effects of
Training Programs on Seven Dependent Variables

Variable	Coworker		Consultant		Self		ANOVA	
	M	SD	M	SD	M	SD	$F(2, 297)$	η^2
Job performance	12.34	2.89	11.78	3.45	10.90	2.98	35.87***	.19
Organizational								
commitment	9.54	1.51	9.67	1.47	9.89	1.32	4.43*	.03
Job commitment	3.35	0.89	3.41	0.96	3.33	0.82	1.67	.01
Job satisfaction	5.67	1.01	4.79	0.99	3.45	1.10	6.78**	.04
Turnover intention	1.44	0.56	1.89	0.67	2.02	0.59	7.42***	.05
Job stress	15.87	3.56	15.32	3.24	17.04	3.18	9.01***	.06
Role ambiguity	4.45	1.32	4.39	4.04	1.25	1.35	3.21*	.02

Note. η^2 = effect size.

*p < .05. **p < .01. ***p < .001.

Table 3.6

In this sample table, ANOVA results have been labeled. Effect sizes could be included as well.

Table X

Means, Standard Deviations, and One-Way Analyses of Variance (ANOVAs) for Effects of
Coworker, Consultant, and Self Training Programs on Seven Dependent Variables

Variable	Coworker		Consultant		Self		ANOVA	
	M	SD	M	SD	M	SD	$F(2, 297)$	p
Job performance	12.34	2.89	11.78	3.45	10.90	2.98	35.87	.001
Organizational								
commitment	9.54	1.51	9.67	1.47	9.89	1.32	4.43	.05
Job commitment	3.35	0.89	3.41	0.96	3.33	0.82	1.67	ns

(Table X continues)

(Table X continued)

Variable	Coworker		Consultant		Self		ANOVA	
	M	SD	M	SD	M	SD	F(2, 297)	p
Job satisfaction	5.67	1.01	4.79	0.99	3.45	1.10	6.78	.01
Turnover intention	1.44	0.56	1.89	0.67	2.02	0.59	7.42	.001
Job stress	15.87	3.56	15.32	3.24	17.04	3.18	9.01	.001
Role ambiguity	4.45	1.32	4.39	4.04	1.25	1.35	3.21	.05

> American Psychological Association (APA) style prefers asterisk notes for expressing significance levels. APA style recommends columns only when there are five or more different *p* values.

Factorial Analysis of Variance: What Is It?

The factorial ANOVA is similar to the one-way ANOVA except there are two or more independent variables. The effects of the independent variables on a single dependent variable are examined.

What Tables Are Used?

Two types of tables usually are used: (a) a table that identifies the means and standard deviations for each cell of the design (see Tables 3.7 and 3.15) and (b) an ANOVA summary table (see Tables 3.8, 3.9, 3.11, 3.12, 3.13, 3.14, and 3.16). Sometimes descriptive statistics and ANOVA results are presented together (see Table 3.10).

What information the writer wishes to present in the ANOVA table (e.g., degrees of freedom, sums of squares, mean squares, *F* ratios, or significance levels) depends on space availability, how comprehensive he or she wishes to be, and whether the results of more than one ANOVA are to be presented in the same table. Examples 3 and 5 provide sample tables for the results of a single factorial ANOVA (Tables 3.7, 3.8, 3.9 and Tables 3.15, 3.16, respectively). Example 4 provides sample tables for the presentation of more than one factorial ANOVA in a single table (Tables 3.10, 3.11, 3.12, 3.13, and 3.14). Note that researchers often will present significant interactions in a figure (this is not illustrated in this book).

"Play It Safe" Table

The most comprehensive table includes the degrees of freedom, sums of squares, mean squares, F ratios, and significance levels for the main effects and interaction effects. Information regarding the within-groups, between-groups, and total sources of variation is included as well (see Table 3.8). If the goal is to be comprehensive in presenting the results, then a table of means and standard deviations also should be included (see Table 3.7).

Example 3

A school would like to see the effects of different classroom and laboratory instructional media on students' grades. A total of 120 students attended a 3-hr psychology class once a week for 3 months. There are three levels of this independent variable (i.e., three different classroom conditions):

1. *Lecture only.* An instructor lectured on material obtained from a textbook, but students were not provided with the textbook.
2. *Textbook only.* Students were provided with only a textbook that they could read during the classroom period.
3. *Multimedia computer-assisted instruction only.* Students learned about the textbook material through a multimedia computer-assisted instructional program during the classroom period; no textbook was provided.

In addition, students attended one of two 1-hr laboratory instruction sessions that took place once a week. The content of the laboratory sessions focused on material presented in the previous class. There are two levels of this independent variable (i.e., two different laboratory conditions):

1. *Group discussion only.* Students were involved in group discussion sessions in which the instructor acted as a facilitator.
2. *Problem solving only.* Students were involved in problem-solving sessions.

Therefore, there are two independent variables (three levels of classroom instruction and two levels of laboratory instruction) and one dependent variable (overall course grade, measured here as a continuous variable).

Table 3.7

Table X

Means and Standard Deviations for Class Conditions as a Function of Laboratory Condition

Class	Group discussion		Problem solving	
	M	SD	M	SD
Lecture	72.45	3.65	71.34	2.45
Textbook	70.31	4.58	71.99	3.81
Multimedia	78.64	3.89	76.97	4.62

> Descriptive statistics such as means and standard deviations often are presented first.

> See chapter 13 for other examples of format for tables of means and standard deviations.

When presenting the ANOVA results, the degrees of freedom, sums of squares, mean squares, *F* ratios, and significance levels often are presented for the different effects (i.e., main effects and interaction effects). The degrees of freedom, sum of squares, and mean squares are presented for the within-groups source of variance. The degrees of freedom and sums of squares are presented for the between-groups and total sources of variation. Table 3.8 is a sample based on the study described in Example 3. To simplify Table 3.8, information regarding the total sources of variation could be omitted, as in Table 3.9.

This table, along with a table of means and standard deviations, is the "Play It Safe" table for a factorial analysis of variance with two independent variables.

Table 3.8

Table X

Summary of Two-Way Analysis of Variance for Class and

Laboratory Conditions

Source	df	SS	MS	F
Class	2	76.66	38.33	6.04**
Laboratory	1	45.27	45.27	7.21**
Class × Laboratory	2	25.24	12.62	2.01
Within cells	114	715.92	6.28	
Total	119	863.09		

**$p < .01$.

Table 3.9

Table X

Two-Way Analysis of Variance for Class and Laboratory

Conditions

Source	df	SS	MS	F
Class	2	76.66	38.33	6.04**
Laboratory	1	45.27	45.27	7.21**
Class × Laboratory	2	25.24	12.62	2.01
Residual	114	715.92	6.28	

**$p < .01$.

The total source of variation has not been included in this table. Residual = within-cells source of variation.

When several factorial ANOVAs have been conducted, the researcher can make a separate ANOVA summary table for each ANOVA, such as Table 3.9 (this may be recommended for theses). Otherwise, including all of the analyses in a summary table such as Tables 3.10 through 3.14, which do not contain as much information, is an alternative.

Example 4

As in Example 3, there are two independent variables: three levels of classroom instruction and two levels of laboratory instruction. Additional dependent variables have been included: overall course grade, recall and recognition score, problem-solving score, and course satisfaction ratings. Separate analyses have been conducted for each.

The researcher may choose not to include the error term (i.e., within groups) in the table. In some instances, degrees of freedom may not be included within a table. For example, including the degrees of freedom may occupy too much space in the table, or the degrees of freedom may be the same for each set of independent ANOVA analyses and therefore are redundant if repeated within a table column. In such cases, the degrees of freedom may be identified in a lettered table note (see Table 3.12 for an example).

Exhibit 3.4

Independent Variables
1. Classroom instruction (lecture, textbook, or multimedia computer-assisted instruction)
2. Laboratory instruction (group discussion or problem solving)

Dependent Variables
1. Overall course grade
2. Recall and recognition score
3. Problem-solving score
4. Course satisfaction ratings

Descriptive statistics and ANOVA results are presented together in this table.

Table 3.10

Table X

Means, Standard Deviations, and Analysis of Variance (ANOVA) Results for Class Conditions
as a Function of Laboratory Condition

Class	Group discussion		Problem solving		ANOVA F		
	M	SD	M	SD	Class (C)	Laboratory (L)	C × L
Overall course grade					6.10**	7.21**	2.01
Lecture	72.67	3.89	73.87	2.52			
Textbook	70.52	2.12	71.97	2.45			
Multimedia	76.89	4.01	79.76	3.43			
Recall and							
recognition score					1.70	1.99	1.79
Lecture	25.34	3.45	24.36	2.33			
Textbook	24.98	2.34	25.01	3.12			
Multimedia	24.67	3.02	25.27	2.86			
Problem-solving score					3.62*	4.73*	0.39
Lecture	15.67	1.89	18.34	2.04			
Textbook	14.88	2.31	16.78	1.98			
Multimedia	16.78	2.45	17.12	3.06			
Course satisfaction							
ratings					5.95**	4.67*	4.15*
Lecture	65.43	9.22	45.28	8.75			
Textbook	59.62	9.87	48.36	7.56			
Multimedia	55.32	8.62	70.21	9.91			

*p < .05. **p < .01.

Only ANOVA results are included in this table.

Table 3.11

Table X

Two-Way Analyses of Variance for Overall Grade, Recall and
Recognition and Problem-Solving Scores, and Course
Satisfaction Ratings

Source	df	MS	F
Overall course grade			
Classroom	2	38.33	6.10**
Laboratory	1	45.27	7.21**
Classroom × Laboratory	2	12.62	2.01
Error	114	6.28	
Recall and recognition score			
Classroom	2	10.43	1.70
Laboratory	1	12.26	1.99
Classroom × Laboratory	2	11.01	1.79
Error	114	6.15	
Problem-solving score			
Classroom	2	19.71	3.62*
Laboratory	1	25.78	4.73*
Classroom × Laboratory	2	2.13	0.39
Error	114	5.44	
Course satisfaction ratings			
Classroom	2	51.16	5.95**
Laboratory	1	40.20	4.67*
Classroom × Laboratory	2	35.67	4.15*
Error	114	8.60	

*p < .05. **p < .01.

Table 3.12

Table X

Two-Way Analyses of Variance for the Overall Grade, Recall and

Recognition and Problem-Solving Scores, and Course Satisfaction

Ratings

Variable and source	MS	F	η^2
Overall grade			
Classroom[a]	38.33	6.10**	.09
Laboratory[b]	45.27	7.21**	.05
Classroom × Laboratory[a]	12.62	2.01	.03
Recall and recognition score			
Classroom[a]	10.43	1.70	.03
Laboratory[b]	12.26	1.99	.02
Classroom × Laboratory[a]	11.01	1.79	.03
Problem-solving score			
Classroom[a]	19.71	3.62*	.06
Laboratory[b]	25.78	4.73*	.04
Classroom × Laboratory[a]	2.13	0.39	.01
Course satisfaction ratings			
Classroom[a]	51.16	5.95**	.09
Laboratory[b]	40.20	4.67*	.03
Classroom × Laboratory[a]	35.67	4.15*	.06

Note. η^2 = effect size.

[a] df = 2, 114. [b] df = 1, 114.

*p < .05. **p < .01.

Another way to present the same ANOVA results is shown in Table 3.13. In Table 3.13 the *F* ratios and significance levels are presented for the main and interaction effects for each ANOVA. Because the degrees of freedom are the same for each of the dependent variables presented, they easily can be listed in a single column. As shown, letters may be used as abbreviations for the independent variables.

Mean squares for main and interaction effects are not included in this table.

Table 3.13

Table X

Analyses of Variance Results for Four Course Outcome Measures

		F			
Source	df	Overall course grade	Recall and recognition score	Problem- solving score	Course satisfaction ratings
Classroom (C)	2	6.10*	1.70	3.62*	5.95**
Laboratory (L)	1	7.21**	1.99	4.73*	4.67*
C × L	2	2.01	1.79	0.39	4.15*
S/CL	114	(6.28)	(6.15)	(5.44)	(8.60)

Note. Values in parentheses represent mean square errors. S/CL = within-cells variance.

*p < .05. **p < .01.

Table 3.14

Table X

Two-Way Analyses of Variance for Overall Course Grade, Recall and Recognition and Problem-Solving Scores, and Course Satisfaction Ratings

	F			
Source	Overall course grade ($\underline{N}$ = 120)	Recall and recognition score ($\underline{N}$ = 118)	Problem-solving score ($\underline{N}$ = 119)	Course satisfaction ratings ($\underline{N}$ = 120)
Classroom (three levels)	6.10*	1.70	3.62*	5.95**
Laboratory (two levels)	7.21**	1.99	4.73*	4.67*
Classroom × Laboratory	2.01	1.79	0.39	4.15*
Within-cells variance	(6.28)	(6.15)	(5.44)	(8.60)

Note. Values in parentheses represent mean square errors.

*$\underline{p}$ < .05. **$\underline{p}$ < .01.

Example 5

In this example there are three independent variables. There are three levels of classroom instruction, two levels of laboratory instruction, and an additional independent variable: two levels of course content. The two levels of course content differ as to the specific material that is to be taught. Students are randomly assigned to either an introductory calculus course or an introductory psychology course. With the increase in the number of independent variables in the analyses, the table of means and standard deviations becomes more complex. The means and standard deviations of a 3 × 2 × 2 design are presented in Table 3.15. The dependent variable is overall course grade.

Sometimes the number of participants per cell is not equal. When this is the case, it may be helpful to include the number of participants per cell in the table of means and standard deviations. See chapter 13 for examples.

An ANOVA table for more than two independent variables looks similar to a two-way ANOVA table, except that one additional main effect and three more interaction effects are included. The more independent variables there are, the more main effects and interaction effects there are to be presented (there are $2^k - 1$ effects, where k is the number of independent variables). (See Table 3.16 for an example.) There are several ways to present ANOVA results for more than two independent variables. (See Tables 3.8 and 3.9 for alternate presentations.)

See chapter 13 for other examples of format for tables of means and standard deviations.

Table 3.15

Table X

Means and Standard Deviations for Class Conditions as a

Function of Laboratory Condition and Course Content

Class	Group discussion		Problem solving	
	M	SD	M	SD
Introductory calculus				
Lecture	72.32	2.78	71.90	3.87
Textbook	76.55	3.12	77.43	2.41
Multimedia	69.89	3.65	70.34	2.45

(Table X continues)

(Table X continued)

Class	Group discussion		Problem solving	
	M	SD	M	SD
Introductory psychology				
Lecture	72.67	3.89	73.87	2.52
Textbook	70.52	2.12	71.97	2.45
Multimedia	76.89	4.01	79.76	3.43

Table 3.16

Table X

Analysis of Variance Results for Main Effects and Interaction

Effects of Classroom, Laboratory, and Course Instruction on

Grade

Variable	df	MS	F
Main effect of classroom (CL)	2	666.22	3.42*
Main effect of laboratory (L)	1	664.29	3.41
Main effect of course (CO)	1	588.30	3.02
CL × L	2	533.75	2.74
CL × CO	2	385.70	1.98
L × CO	1	638.94	3.28
CL × L × CO	2	239.60	1.23
Within-cells error	228	194.74	

*p < .05.

Effect sizes could be presented in a column to the right of the F ratios.

Within-Subjects, Mixed, and Hierarchical Designs: What Is It?

Within-subjects, mixed, and hierarchical designs are broad categories that include repeated measures designs, split-plot designs, nested designs, and numerous others that consist of variations of between-subjects and within-subjects designs.

What Tables Are Used?

Two types of tables usually are used: (a) a table that identifies the means and standard deviations for each cell of the design and (b) an ANOVA summary table. The ANOVA summary table could include the degrees of freedom, sums of squares, mean squares, F ratios, and significance levels for the sources, or some of this information could be excluded. (See the factorial ANOVA section for examples.) Also, depending on the complexity of the analysis, a table may not be required; it may be possible to simply present the results within the text.

Example 6 is a mixed design, with one between-subjects variable and one within-subjects variable. The sample table presented (Table 3.18) can easily be adapted to more complex mixed designs, such as those with two between-subjects variables and one within-subjects variable (more main effects and interaction effects would have to be added in rows for the between-subjects and within-subjects sources of variance). For hierarchical designs, if the hierarchical design includes a within-groups variable, then the ANOVA table would look similar to Table 3.18 (with the between-subjects sources of variance, degrees of freedom, sums of squares, mean squares, F ratios, and significance levels shown first followed by the within-subjects sources of variance). Other hierarchical designs should be organized in such a manner that the sources of variance (i.e., main effects, interaction effects, nested, and error sources) are organized in a logical manner. The table of the mixed design (Table 3.18) and the tables of factorial between-subjects designs presented earlier in this chapter can be used as guidelines.

"Play It Safe" Table

There are two "Play It Safe" tables in this section. Table 3.17 includes means and standard deviations, and Table 3.18 includes the degrees of freedom, sums of squares, mean squares, F ratios, and significance levels for the different sources. Tables 3.17 and 3.18 are the "Play It Safe" tables for a mixed design with one between-subjects and one within-subjects variable.

Example 6

Researchers wished to determine the long-term effects of two new anxiety-reducing drugs on individuals who experienced major work-related anxiety for 1 year. There are four levels of this between-subjects variable (i.e., four different medication conditions), with 10 people in each condition:

1. *Drug A.* A new anxiety-reducing drug.
2. *Drug B.* A second new anxiety-reducing drug.
3. *Drug C.* An anxiety-reducing drug currently being prescribed by most physicians.
4. *Drug D.* A placebo.

Each participant's anxiety was measured using a 20-item work-related anxiety questionnaire on several occasions. There are five levels of this within-subjects variable:

1. *Time 1.* Tested immediately after taking the medication.
2. *Time 2.* Tested 1 month after taking the medication.
3. *Time 3.* Tested 3 months after taking the medication.
4. *Time 4.* Tested 6 months after taking the medication.
5. *Time 5.* Tested 1 year after taking the medication.

Thus, there are two independent variables (medication and testing time). The dependent variable is anxiety as measured using a 20-item work-related anxiety questionnaire.

Exhibit 3.6

Independent Variables
1. Medication (Drug A, B, C, or D)
2. Testing time (Times 1, 2, 3, 4, and 5)

Dependent Variable
1. Anxiety

The means and standard deviations for each cell often are presented first.

Table 3.17

Table X

Means and Standard Deviations for Four Drugs and Five
Testing Times

Testing time	Drug A	Drug B	Drug C	Drug D
Time 1				
M	121.24	116.34	114.29	120.87
SD	20.65	21.67	19.90	17.89
Time 2				
M	119.31	105.12	100.78	40.35
SD	19.77	20.82	22.54	15.43
Time 3				
M	60.75	110.96	90.27	35.21
SD	10.64	21.47	17.89	9.75
Time 4				
M	45.76	95.31	41.18	42.85
SD	11.56	15.87	9.87	9.45
Time 5				
M	41.21	112.76	44.55	38.34
SD	9.87	18.79	10.88	9.89

See chapter 13 for other examples of format for tables of means and standard deviations.

> This table, along with a table of means and standard deviations, is the "Play-It-Safe" table for a mixed design with one between-subjects and one within-subjects variable.

Table 3.18

Table X

Analysis of Variance Results for Medication and Time Variables

Source	df	SS	MS	F
Between subjects				
Drug	3	10,146.01	3,382.00	12.17*
Error 1	36	10,004.25	277.90	
Within subjects				
Time	4	5,076.48	1,269.12	45.83*
Drug × Time	12	2,194.26	182.86	6.60*
Error 2	144	3,987.90	27.69	

*p < .05.

> Effect sizes could be presented in a column to the right of the *F* ratios.

CHAPTER 4

Canonical Correlation

What Is It?

The canonical correlation is used to measure the relationships between two sets of variables.

What Tables Are Used?

The results of canonical correlations usually are reported in one table that includes the correlations and standardized canonical coefficients between the sets of variables and their canonical variates (see Table 4.1 or 4.2). Note that the variates sometimes are referred to as *roots*.

"Play It Safe" Table

The "Play It Safe" choice for the format of the canonical correlation table is Table 4.1 because it is most comprehensive.

Example

The sample data used here are from a study in which the researchers are interested in the relationship between two sets of variables. One set of variables measures job satisfaction, and the other set measures participants' personal characteristics. The job satisfaction variables include an overall satisfaction rating, satisfaction with working conditions, satisfaction with amount of work, and satisfaction with promotion prospects. The personal characteristics variables include education, health, income, and age. The researchers want to determine how these two sets of variables are related for a sample of employees at a major multinational corporation.

The researchers should report the significant canonical correlations in the text. For this example, there are two pairs of canonical variates that account for significant relationships between the two sets of variables. These are the canonical variates that should be presented in the table.

Exhibit 4.1

Variables

1. Job satisfaction (overall satisfaction rating, satisfaction with working conditions, satisfaction with amount of work, satisfaction with promotion prospects)
2. Personal characteristics (education, health, income, age)

> This is the "Play It Safe" table for canonical correlation results.

Table 4.1

Table X

<u>Correlations and Standardized Canonical Coefficients Between Job Satisfaction and</u>

<u>Personal Characteristics Variables and Their Canonical Variates</u>

	First variate		Second variate	
Variable	Correlation	Canonical coefficient	Correlation	Canonical coefficient
Job satisfaction				
Overall satisfaction rating	.72	.66	.25	.59
Working conditions	−.56	−.34	−.31	−.44
Amount of work	.78	.44	.12	.23
Promotion prospects	.19	.12	.45	.34
Personal characteristics				
Education	.87	.76	.45	.23
Health	.67	.52	.21	.11
Income	.92	.69	.59	.54
Age	−.65	−.55	−.33	−.08

Table 4.2

Table X

Canonical Analysis of Job Satisfaction and Personal

Characteristics Variables

	Standardized canonical coefficient	
Variable	Root 1	Root 2
Job satisfaction		
Overall satisfaction rating	.66	.59
Working conditions	-.34	-.44
Amount of work	.44	.23
Promotion prospects	.12	.34
Personal characteristics		
Education	.76	.23
Health	.52	.11
Income	.69	.54
Age	-.55	-.08

CHAPTER 5

Chi-Square

What Is It?

The chi-square is used to determine whether differences between observed and expected frequencies are statistically significant.

What Tables Are Used?

The tables that are commonly used to present the results of chi-square analyses are frequency tables with a column for chi-square values to indicate whether certain frequencies are significantly different from each other (Tables 5.1 and 5.2).

"Play It Safe" Table

Tables 5.1 and 5.2 are both comprehensive, so either could be considered a "safe" choice. However, Table 5.1 uses the American Psychological Association's (APA's) preferred style for presenting significance levels.

Example

In this study, the researchers are investigating certain health problems in infancy. They are interested in whether there is any sex difference in the rate of occurrence of these health problems. Their data come from the health records of the first year of life for 266 infants who received regular medical care and immunizations. The independent variable is the type of health problem: eye infections, ear infections, strep throat infections, upper respiratory viruses, pneumonia, and bronchitis. The dependent variable is the frequency of occurrence of these illnesses.

Exhibit 5.1

Independent Variables
1. Health problems (eye infections, ear infections, strep throat infections, upper respiratory viruses, pneumonia, bronchitis)
2. Sex (boys vs. girls)

Dependent Variable
1. Frequency of occurrence of illness (occurrence vs. nonoccurrence)

Table 5.1

Note that according to APA publication language style guidelines, participants should not be referred to as *males* and *females* because *male* and *female* are adjectives.

Table X

Prevalence (%) of Six Illnesses Among Male and Female Infants
in the First Year of Life

Illness	Boys (n = 123)	Girls (n = 143)	$\chi^2(1)$
Eye infections	21	14	12.81***
Ear infections	44	29	21.34***
Strep throat infections	19	18	1.26
Upper respiratory viruses	89	68	29.43***
Pneumonia	31	31	0.92
Bronchitis	13	6	11.18***

***p < .001.

This is the "Play It Safe" table.

This table presents the same information as Table 5.1 but uses a different format for presenting significance levels.

Table 5.2

Table X

Prevalence (%) of Six Illnesses Among Male and Female Infants in the First Year of Life

Illness	Boys ($\underline{n}$ = 123)	Girls ($\underline{n}$ = 143)	$\chi^2(1)$	$\underline{p}$
Eye infections	21	14	12.81	.001
Ear infections	44	29	21.34	.001
Strep throat infections	19	18	1.26	$\underline{ns}$
Upper respiratory viruses	89	68	29.43	.001
Pneumonia	31	31	0.92	$\underline{ns}$
Bronchitis	13	6	11.18	.001

According to APA style, asterisk notes are preferred for significance levels; columns are recommended only when there are five or more *p* values.

CHAPTER 6

Cluster Analysis

What Is It?

Cluster analysis refers to a variety of techniques used to determine the underlying structure, natural grouping, or conceptual scheme of a set of entities by illustrating which of those entities are most closely related based on a set of descriptors (e.g., attitudes, interests, symptoms, or traits). The underlying structure or natural groupings often are referred to as *clusters*. There are various methods (e.g., agglomerative hierarchical clustering and divisive hierarchical clustering) and measures (e.g., square Euclidean distance and Pearson product–moment correlation) for calculating distances between descriptors. And there are various methods (e.g., between-groups linkage, nearest neighbor, and farthest neighbor) for combining descriptors into clusters. Agglomerative hierarchical clustering is illustrated in this chapter. The squared Euclidean distance is the measure of distance, and the between-groups linkage is the method used to determine whether two entities should be combined.

What Tables Are Used?

The results of cluster analyses are never presented in a table. The various clusters generally are presented in either a dendrogram or a figure. If the researcher has numerous descriptors and wishes to determine whether a combination of these descriptors form different clusters, a dendrogram is used. If the researcher wishes to determine whether profiles can be created for individuals (or objects or animals) based on a set of variables, then a figure is used. A dendrogram is illustrated in Example 1 (Figure 6.1), and a figure is presented in Example 2 (Figure 6.2).

In addition to dendrograms or figures, a table of means and standard deviations for all of the variables should be included (see Table 6.1). If one purpose is to examine differences between profiles of individuals on the variables that make up those profiles, then a table of means and standard deviations and the results of any tests of group differences that may have been conducted (e.g., *F* tests, *t* tests, or post hoc analyses) are also presented (see Table 6.2).

"Play It Safe" Table

The most comprehensive table is one including the means and standard deviations and the results of any tests of group differences conducted between the clusters (see Table 6.2). In addition, a dendrogram or figure should be included to illustrate the results of the cluster analysis (see Figures 6.1 and 6.2, respectively).

Example 1

Two researchers have videotaped 192 individuals. Each videotape consists of a person having a conversation with a close friend about what constitutes a liberal education. The researchers have coded numerous behaviors such as irrelevant hand gestures, smiling, eye contact, and so on. A total of 10 behaviors were coded. The researchers wish to determine whether an individual's interpersonal communication style could be described by clusters of these 10 behaviors.

Exhibit 6.1

Variables
1. Establishes eye contact
2. Listens attentively
3. Relates topic to other person
4. Interrupts frequently
5. Relates topic only to self
6. Asks questions
7. Does not answer questions
8. Uses irrelevant hand gestures
9. Speed of speech
10. Smiles

This table presents the means and standard deviations for each variable. For more example formats of tables of means and standard deviations, see chapter 13.

Table 6.1

Table X

Descriptive Statistics for 10 Communication Behaviors (N = 192)

Communication behaviors	$\underline{M}$	$\underline{SD}$
Establishes eye contact	28.96	21.59
Listens attentively	29.77	19.17
Relates topic to other person	27.39	24.46
Interrupts frequently	31.33	19.87
Relates topic only to self	30.39	21.80
Asks questions	30.22	21.44
Does not answer questions	30.04	22.22
Uses irrelevant hand gestures	29.93	21.61
Speed of speech	30.87	21.90
Smiles	36.13	23.58

Figure 6.1

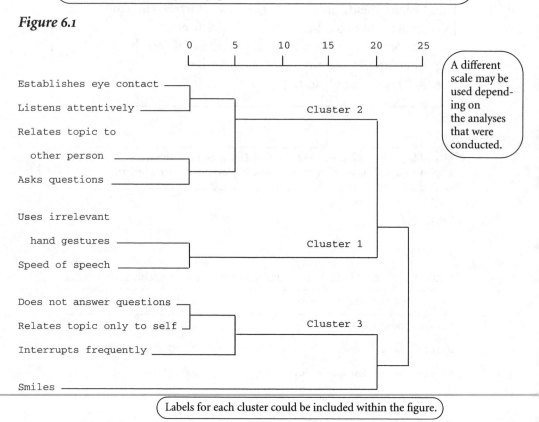

A different scale may be used depending on the analyses that were conducted.

Figure X. Dendrogram based on the results of the agglomerative cluster analysis.

Example 2

Two researchers wish to determine whether profiles exist for various communication styles. They videotaped 88 participants having a conversation with a close friend about the physiological, social, and economic benefits of being a vegetarian. The researchers obtained three measures of communication: one physical, one emotional, and one conversational. They wished to determine whether communication profiles can be formed using these three measures and, if so, what form these profiles would take.

This is a "Play It Safe" figure for the results of a cluster analysis. Table 6.2 is also required to present the differences among the various profiles.

According to APA format, figure captions should be listed together on a separate page.

Figure 6.2

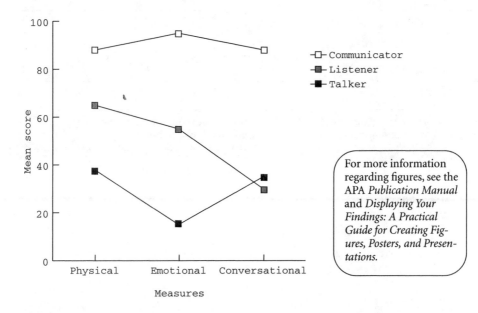

For more information regarding figures, see the APA *Publication Manual* and *Displaying Your Findings: A Practical Guide for Creating Figures, Posters, and Presentations.*

<u>Figure X</u>. Mean-score profiles for three types of communication styles.

This is a "Play It Safe" table for the results of a cluster analysis.

If different analyses were conducted to determine group differences, they would be presented in the column in which F ratios are located. Alternatively, some researchers prefer to place this information in the results section of the text rather than in a table.

Table 6.2

Table X

Between-Groups Differences for Physical, Emotional, and Conversational Measures

	Group 1 (communicators) (n = 18)		Group 2 (listeners) (n = 16)		Group 3 (talkers) (n = 54)		
Measure	M	SD	M	SD	M	SD	$F(2, 85)$
Physical	89.28	7.43	64.50	6.09	37.00	4.12	703.50***
Emotional	95.28	4.07	54.75	9.42	14.83	4.18	1,548.85***
Conversational	90.28	4.86	27.50	5.77	34.00	4.36	1,074.86***

***p < .001.

For more examples of how to present F ratios, the results of t tests, and the results of other post hoc analyses, see chapters 3, 20, and 18, respectively.

CHAPTER 7

Correlation

What Is It?

A correlation is a measure of the direction and magnitude of the linear relation between two variables.

What Tables Are Used?

When reporting the results of correlational analyses, a table usually is necessary only if there are more than two variables. In that case, the relevant tables are a table of means and standard deviations (Table 7.1), and a table of intercorrelations between all variables included in the analysis (Tables 7.2, 7.3, or 7.4), or a table of correlations between two sets of variables (Table 7.6). The table of means and standard deviations and the table of intercorrelations can be combined into one table (Tables 7.5 or 7.6). The table format used depends on the nature of the study.

"Play It Safe" Table

All of the tables in this chapter could be considered "safe" choices because they all present the same amount of information about the correlations obtained from the data. Thus, no one table is any more or less comprehensive than any other.

Example 1

In the study presented here the researchers have constructed a new scale to measure need for achievement. They call their new scale the Dimensions of Achievement Scale (DAS). To examine how the DAS relates to existing measures of need for achievement, the researchers want to correlate scores on the DAS with scores on other measures. They have 100 participants complete each of the six measures. Their variables are the different measures of need for achievement: the DAS, the Brunswick Achievement Measure, the Need for Achievement Inventory, the Achievement Perception Test, a peer rating of need for achievement, and a self-rating of need for achievement.

Exhibit 7.1

Variables

1. Dimensions of Achievement Scale
2. Brunswick Achievement Measure
3. Need for Achievement Inventory
4. Achievement Perception Test
5. Peer rating of need for achievement
6. Self-rating of need for achievement

See chapter 13 for other examples of format for tables of means and standard deviations.

Table 7.1

Table X

Means and Standard Deviations for Six Measures of Need for

Achievement

Measure	M	SD
Dimensions of Achievement Scale	43.21	14.34
Brunswick Achievement Measure	22.22	8.75
Need for Achievement Inventory	12.15	3.47
Achievement Perception Test	14.09	5.37
Peer rating of need for achievement	12.30	5.57
Self-rating of need for achievement	11.91	4.91

This is the most common format for a table of intercorrelations.

Table 7.2

Table X

Intercorrelations for Dimensions of Achievement Scale and Five Other

Need-for-Achievement Measures

Measure	1	2	3	4	5	6
1. Dimensions of Achievement Scale	--					
2. Brunswick Achievement Measure	.76	--				
3. Need for Achievement Inventory	.70	.88	--			
4. Achievement Perception Test	.56	.65	.61	--		
5. Peer rating of need for achievement	.45	.55	.52	.67	--	
6. Self-rating of need for achievement	.53	.56	.43	.37	.87	--

Note. All coefficients are significant at $p < .01$.

> In this table, the data for two groups of participants are presented—one above the diagonal and one below.

Table 7.3

Table X

Intercorrelations for Scores on Six Measures of Need for Achievement as a Function of Gender

Measure	1	2	3	4	5	6
1. DAS	--	.86	.76	.60	.43	.63
2. BAM	.66	--	.80	.70	.55	.50
3. NAchI	.64	.96	--	.62	.52	.40
4. APT	.52	.60	.60	--	.77	.37
5. Peer	.47	.55	.52	.57	--	.90
6. Self	.43	.61	.45	.37	.85	--

Note. Intercorrelations for male participants (n = 50) are presented above the diagonal, and intercorrelations for female participants (n = 50) are presented below the diagonal. All coefficients are significant at p < .01. DAS = Dimensions of Achievement Scale; BAM = Brunswick Achievement Measure; NAchI = Need for Achievement Inventory; APT = Achievement Perception Test; Peer = peer rating of need for achievement; Self = self-rating of need for achievement.

When relevant, researchers sometimes present coefficient alphas for the different measures in a correlation table, as illustrated in this table.

Table 7.4

Table X

Intercorrelations and Coefficient Alphas for Scores on Six

Measures of Need for Achievement

Measure	1	2	3	4	5	6
1. DAS	**.91**					
2. BAM	.76	**.89**				
3. NAchI	.70	.88	**.92**			
4. APT	.56	.65	.61	**.84**		
5. Peer	.45	.55	.52	.67	**.78**	
6. Self	.53	.56	.43	.37	.87	**.80**

Note. Coefficient alphas are presented in boldface along the diagonal. All coefficients are significant at p < .01. DAS = Dimensions of Achievement Scale; BAM = Brunswick Achievement Measure; NAchI = Need for Achievement Inventory; APT = Achievement Perception Test; Peer = peer rating of need for achievement; Self = self-rating of need for achievement.

> Instead of using a separate table of means and standard deviations, these values can be included in a correlation table.

Table 7.5

Table X

Intercorrelations, Means, and Standard Deviations for Scores on
Six Measures of Need for Achievement

Measure	BAM	NAchI	APT	Peer	Self	M	SD
DAS	.76	.70	.56	.45	.53	43.2	14.3
BAM		.88	.65	.55	.56	22.2	8.8
NAchI			.61	.52	.43	12.2	3.5
APT				.67	.37	14.1	5.4
Peer					.87	12.3	5.6
Self						11.9	4.9

Note. All coefficients are significant at $p < .01$. DAS = Dimensions of Achievement Scale; BAM = Brunswick Achievement Measure;
NAchI = Need for Achievement Inventory; APT = Achievement Perception Test; Peer = peer rating of need for achievement; Self =
self-rating of need for achievement.

Example 2

There are six subscales of the DAS. Each subscale measures a different domain in which need for achievement might be manifested: School, Family Relationships, Friendships, Work, Athletics, and Hobbies. The researchers want to determine how each of the DAS subscales correlates with the other measures of achievement included in Example 1 (see Table 7.6).

Table 7.6

Table X

Means, Standard Deviations, and Correlations of Dimensions of Achievement Scale Subscales With Measures of Need for Achievement

DAS subscale	M	SD	Measure				
			BAM	NAchI	APT	Peer	Self
School	22.3	7.7	.76**	.89**	.45*	.34*	.55*
Family							
Relationships	20.9	6.7	.81**	.79**	.73**	.44*	.20
Friendships	16.8	6.1	.77**	.82**	.23	.39*	.45*
Work	24.6	6.9	.80**	.79**	.45*	.41*	.63*
Athletics	22.5	9.8	.66*	.77**	.19	.75**	.21
Hobbies	14.3	9.1	.79**	.88**	.56*	.21	.22

Note. DAS = Dimensions of Achievement Scale; BAM = Brunswick Achievement Measure; NAchI = Need for Achievement Inventory; APT = Achievement Perception Test; Peer = peer rating of need for achievement; Self = self-rating of need for achievement.

*p < .05. **p < .01.

Means and standard deviations could be presented in a separate table. See chapter 13 for other examples of format for tables of means and standard deviations.

CHAPTER 8

Discriminant Function Analysis

What Is It?

Discriminant function analysis allows prediction of group membership (when groups are different levels of the dependent variable) from a set of predictor variables.

What Tables Are Used?

If the analysis is a stepwise discriminant function analysis, the four tables most commonly used to present results are (a) a table of means and standard deviations for predictor variables as a function of group (Table 8.1), (b) a table of discriminant function results with Wilks's lambda results for each step (Table 8.2), (c) a table of discriminant function coefficients (Table 8.3 or 8.4), and (d) a table of results of classification analysis for the discriminant function analysis (Table 8.5 or 8.6). Occasionally, there is also a centroids plot if there are three or more groups (levels of the dependent variable) involved in the analysis (Figure 8.1).

If the predictor variables in the discriminant function analysis were not entered stepwise, then the tables used are the same with one exception: The table for Wilks's lambda results usually is omitted, and the Wilks's lambda results are presented within the text instead of in a table.

"Play It Safe" Table

The "Play It Safe" tables for a stepwise discriminant function analysis are Tables 8.1, 8.2, 8.3, and 8.5. The "Play It Safe" tables for other discriminant function analyses are Tables 8.1, 8.3, and 8.5. The "safe" choice for stepwise and all other discriminant function analyses also includes a centroids plot (Figure 8.1), particularly if there are three or more groups (levels of the dependent variable).

Example

In this study, the researchers are interested in the factors that can predict chess skill (the dependent variable). They recruited three groups of participants: chess experts, chess amateurs, and chess novices. There were 30 participants in each group. The researchers measure participants' performance on several predictor variables to determine whether performance on those variables can predict group membership. The predictor (independent) variables are spatial ability, problem-solving ability, map-reading skill, and accuracy of visual imagery.

Exhibit 8.1

Independent Variables
  1. Spatial ability
  2. Problem-solving ability
  3. Map-reading skill
  4. Accuracy of visual imagery

Dependent Variable
  1. Chess skill

This is a "Play It Safe" table for all types of discriminant function analyses.

For other examples of tables of means and standard deviations and tables of a priori contrasts, see chapters 13 and 18, respectively.

Table 8.1

Table X

Means and Standard Deviations of Predictor Variables as a Function of Chess Skill

Predictor variable	Experts		Amateurs		Novices	
	M	SD	M	SD	M	SD
Spatial ability	$16.60_{a,b}$	2.67	$12.56_{a,c}$	1.92	$7.90_{c,b}$	1.99
Problem-solving ability	$89.60_{d,e}$	6.17	73.17_d	7.61	67.40_e	10.34
Map-reading skill	$15.50_{f,g}$	2.67	12.23_f	2.86	13.56_g	3.27
Accuracy of visual imagery	37.73	4.36	34.37	6.96	33.70	10.85

Note. Means with the same subscript differ significantly at $p < .01$.

This is a "Play It Safe" table for a stepwise discriminant function analysis. Information for other discriminant function analyses could be presented in the text.

Table 8.2

Table X

Predictor Variables in Stepwise Discriminant Function Analysis

Step	Predictor variable	Variables in discriminant function	Wilks's λ	Equivalent $F(2, 167)$
1	Spatial ability	1	.274	115.25***
2	Problem-solving ability	2	.175	59.70***
3	Map-reading skill	3	.157	43.26***
4	Accuracy of visual imagery	4	.142	34.80***

***$p < .001$.

This is a "Play It Safe" table for all types of discriminant function analyses.

Table 8.3

Table X

Correlation of Predictor Variables With Discriminant Functions (Function
Structure Matrix) and Standardized Discriminant Function Coefficients

	Correlation with discriminant functions		Standardized discriminant function coefficients	
Predictor variable	Function 1	Function 2	Function 1	Function 2
Spatial ability	.735	−.376	.864	−.544
Problem-solving ability	.517	.527	.683	.492
Map-reading skill	.169	.657	−.135	.786
Accuracy of visual imagery	.099	.151	.353	.128

This table illustrates that sometimes a table of coefficients presents only the structure matrix and sometimes coefficients are presented to only two decimal places.

Table 8.4

Table X

Correlations Between Discriminating Variables and Discriminant
Functions (Function Structure Matrix)

Variable	Function 1	Function 2
Spatial ability	.74	−.38
Problem-solving ability	.52	.53
Map-reading skill	.17	.66
Accuracy of visual imagery	.10	.15

This is a "Play It Safe" table for all types of discriminant function analyses.

Table 8.5

Table X

Classification Analysis for Chess Skill

		Predicted group membership					
		Experts		Amateurs		Novices	
Actual group membership	n	n	%	n	%	n	%
Experts	30	29	96.7	1	3.3	0	0.0
Amateurs	30	2	6.7	26	86.7	2	6.7
Novices	30	0	0.0	2	6.7	28	93.3

Note. Overall percentage of correctly classified cases = 92.2%.

Table 8.6

Table X

Classification Analysis for Chess Skill

		Predicted group membership		
Actual group membership	n	Experts	Amateurs	Novices
Experts	30			
n		29	1	0
%		96.7	3.3	0.0
Amateurs	30			
n		2	26	2
%		6.7	86.7	6.7
Novices	30			
n		0	2	28
%		0.0	6.7	93.3

Results of the classification analysis also could be presented in this format.

Note. Overall percentage of correctly classified cases = 92.2%.

Figure 8.1

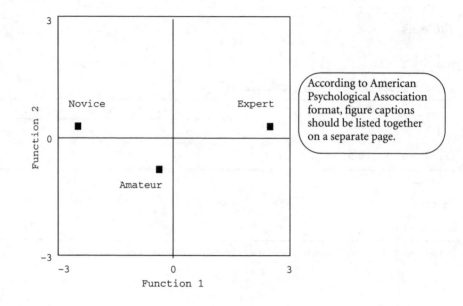

Figure X. Group centroids plot from discriminant function analysis.

CHAPTER 9

Factor Analysis

What Is It?

Factor analysis is a multivariate statistical technique used to reduce the number of observed variables to a smaller number of latent variables, identified as *factors*. Different factor analytic procedures exist. The single term *factor analysis* is used in this chapter to refer to all types of factor analytic procedures.

What Tables Are Used?

Generally, one type of table is used that includes the factor loadings (rotated or unrotated) for the factors of interest. There are many ways these tables can be presented. For example, only the factor loadings may be presented (see Tables 9.4 and 9.11). Additional information regarding the factor analysis results can be included, such as the percentage of variance (total, common, or both), eigenvalues, and communalities (see Tables 9.1, 9.2, 9.3, and 9.10). Sometimes, information such as means, standard deviations, factor scores, and factor correlations is presented along with the factor loadings (see Tables 9.5, 9.6, 9.7, 9.8, and 9.9).

Example 1 presents the results of a single factor analysis without a factor rotation. Examples 2 and 3 present the results of a varimax rotation and an oblimin rotation, respectively. Example 4 presents the results of two separate factor analyses.

"Play It Safe" Table

The "Play It Safe" table would include (a) all of the factor loadings, eigenvalues, communalities, and percentages of variance for an unrotated factor analysis solution (see Table 9.1); (b) all of the factor loadings and communalities for an orthogonally rotated factor analysis solution (see Table 9.3); and (c) all of the factor loadings for an obliquely rotated solution (see Table 9.8). The rotated (orthogonal or oblique) factor analysis solution is more commonly presented in a table for theses and journal articles.

For theses, a more thorough description of the factor analysis may be required such as a scree plot, which is not illustrated here. Additionally, should a factor analysis of a measure be conducted, then it would be best to include descriptions of the items (as in Table 9.3, 9.4, 9.5, and 9.8). If this is not possible, then a description of the items can be included in an appendix (e.g., see Tables 9.1, 9.6, 9.7, 9.9, 9.10, and 9.11). Table 9.10 is an example of the "Play It Safe" table for the results of two separate factor analyses.

Example 1

Elementary school teachers ($N = 81$) were asked to rate the importance of several reasons for elementary school children failing their grade. A newly developed questionnaire, the 15-item Student Failure Questionnaire (SFQ), was provided to all participants. A principal-components factor analysis was conducted to identify the underlying factors of this new questionnaire.

The study consists of 15 variables: Each of the 15 items of the SFQ constitutes one variable, with each item identifying a potential reason for elementary school children failing their grade.

Exhibit 9.1

Variables

Student Failure Questionnaire Items

1. Student is excessively absent from school.
2. Student does extra work on his or her own.
3. Education is taught in a second language.
4. Student asks questions about material.
5. Student lacks adequate diet.
6. Student lacks support from school.
7. Student does not do homework.
8. Student lacks support from teacher.
9. Student has attention deficit disorder.
10. Student lacks support from other students.
11. Student appears to be of superior intellect.
12. Student lacks parental support.
13. Student makes extra effort to participate in class.
14. Student has learning disability.
15. Student enjoys doing schoolwork.

Table 9.1

```
Table X

Factor Loadings From Principal-Components Analysis:

Communalities, Eigenvalues, and Percentages of Variance
```

	Factor loading			
Item	1	2	3	Communality
1	.73	.45	−.46	.95
2	.66	.36	−.53	.85
3	.90	.18	.32	.94
4	.75	−.56	.25	.94
5	.29	−.43	−.34	.38
6	.86	.34	.25	.92
7	.75	.21	−.42	.78
8	.61	.67	−.17	.85
9	.54	−.36	.42	.60

(Table X continues)

This table presents the results of a factor analysis without any factor rotation.

This is the "Play It Safe" table for an unrotated factor analysis solution.

Item descriptions may be presented in an appendix.

(Table X continued)

Item	Factor loading 1	Factor loading 2	Factor loading 3	Communality
10	.57	.61	-.17	.72
11	.76	-.46	.33	.90
12	.78	.21	-.29	.74
13	.33	-.43	.41	.46
14	.84	-.35	.15	.85
15	.79	-.14	.31	.74
Eigenvalues	7.34	2.56	1.72	
% of variance	48.93	17.07	11.47	

This is a summary table for the factors.

Table 9.2

Table X

Eigenvalues, Percentages of Variance, and Cumulative Percentages
for Factors of the 15-Item Student Failure Questionnaire

Factor	Eigenvalue	% of variance	Cumulative %
1	7.34	48.93	48.93
2	2.56	17.07	66.00
3	1.72	11.47	77.47

Occasionally, as shown by this table, a summary table of the eigenvalues, percentages of total variance, and cumulative percentages is used.

Example 2

In this example, the 15-item SFQ, described in Example 1, is subjected to a varimax orthogonal rotation.

This is the "Play It Safe" table for a varimax orthogonally rotated factor analysis solution.

Table 9.3

Table X

Summary of Items and Factor Loadings for Varimax Orthogonal Three-Factor Solution for the Student Failure Questionnaire (N = 81)

Item	Factor loading			Communality
	1	2	3	
3. Education is taught in a second language.	**.89**	.02	−.12	.80
14. Student has learning disability.	**.86**	−.02	.02	.74
9. Student has attention deficit disorder.	**−.86**	.11	.13	.76
5. Student lacks adequate diet.	**.79**	.04	−.14	.65
11. Student appears to be of superior intellect.	**.79**	.12	.03	.64
1. Student is excessively absent from school.	**.77**	−.16	.08	.62
7. Student does not do homework.	.05	**.86**	−.08	.76
15. Student enjoys doing schoolwork.	.01	**.79**	.10	.63
2. Student does extra work on his or her own.	.03	**.77**	.09	.60
13. Student makes extra effort to participate in class.	.10	**.75**	.32	.67
4. Student asks questions about material.	−.09	**.69**	.32	.59
12. Student lacks parental support.	.09	.09	**.90**	.82
6. Student lacks support from school.	.09	.11	**.89**	.81
8. Student lacks support from teacher.	.08	.06	**.75**	.58
10. Student lacks support from other students.	.06	.10	**.62**	.39

Note. Boldface indicates highest factor loadings.

Items are ordered according to their factor loadings (from highest to lowest) and grouped according to factor. However, items could be presented in numerical order as on the questionnaire.

Table 9.4

Table X

Factor Loadings for Varimax Orthogonal Three-Factor Solution

Item	Factor loading
Factor 1: External Reasons (Out of Student's Control)	
3. Education is taught in a second language.	.89
14. Student has learning disability.	.86
9. Student has attention deficit disorder.	−.86
5. Student lacks adequate diet.	.79
11. Student appears to be of superior intellect.	.79
1. Student is excessively absent from school.	.77
Factor 2: Student Effort	
7. Student does not do homework.	.86
15. Student enjoys doing schoolwork.	.79
2. Student does extra work on his or her own.	.77
13. Student makes extra effort to participate in class.	.75
4. Student asks questions about material.	.69
Factor 3: Support From Others	
12. Student lacks parental support.	.90
6. Student lacks support from school.	.89
8. Student lacks support from teacher.	.75
10. Student lacks support from other students.	.62

Note. N = 81 and α = .76 for entire measure.

Only the highest factor loading for each item is presented (i.e., a single factor loading for each item is shown rather than three factor loadings.) In this example, factor loadings greater than .40 are considered high (the specific cutoff used varies according to the researcher).

Sometimes the writer may wish to include additional information in the table. Tables 9.5 through 9.7 provide different examples of how this can be achieved. In Table 9.5 item means for boys and girls within the sample are provided along with the factor loadings. Table 9.6 provides item means and standard deviations for the entire sample, rotated item factor loadings for all three factors, and item communalities. Table 9.7 reports the internal consistency of the items that form a factor and the factor loadings. (Internal consistency information regarding a set of items is useful to determine if the items are to be considered as separate scales. This information can be presented within the text.)

Table 9.5

Table X

Item Means for Respondents and Factor Loadings From Principal-Components Analysis With Varimax Rotation

Item	M Boys	Girls	Factor loading
Factor 1: External Reasons (Out of Student's Control)			
3. Education is taught in a second language.	5.23	5.39	.89
14. Student has learning disability.	6.44	6.70	.86
9. Student has attention deficit disorder.	5.29	4.55	−.86
5. Student lacks adequate diet.	4.81	5.13	.79
11. Student appears to be of superior intellect.	3.73	3.55	.79
1. Student is excessively absent from school.	4.52	3.80	.77
Factor 2: Student Effort			
7. Student does not do homework.	5.27	4.91	.86
15. Student enjoys doing schoolwork.	5.44	4.80	.79
2. Student does extra work on his or her own.	4.56	3.52	.77
13. Student makes extra effort to participate in class.	4.40	4.86	.75
4. Student asks questions about material.	5.01	4.65	.69
Factor 3: Support From Others			
12. Student lacks parental support.	5.64	5.10	.90
6. Student lacks support from school.	3.50	3.04	.89
8. Student lacks support from teacher.	3.41	3.51	.75
10. Student lacks support from other students.	2.70	3.16	.62

Note. N = 81. Item mean scores reflect the following response choices: 1 = strongly disagree, 2 = moderately disagree, 3 = slightly disagree, 4 = neither agree nor disagree, 5 = slightly agree, 6 = moderately agree, and 7 = strongly agree.

Table 9.6

Table X

Means, Standard Deviations, Rotated Factor Loadings, and Communalities for Student Failure Questionnaire Items

Item	M	SD	Factor loadings 1	2	3	h^2
3	5.31	1.61	**.89**	.02	−.12	.80
14	6.57	2.45	**.86**	−.02	.02	.74
9	4.92	1.82	**−.86**	.11	.13	.76
5	4.97	2.17	**.79**	.04	−.14	.65
11	3.64	2.98	**.79**	.12	.03	.64
1	4.16	2.01	**.77**	−.16	.08	.62
7	5.09	1.39	.05	**.86**	−.08	.76
15	5.12	2.67	.01	**.79**	.10	.63
2	4.04	1.28	.03	**.77**	.09	.60
13	4.63	2.91	.10	**.75**	.32	.67
4	4.83	1.30	−.09	**.69**	.32	.59
12	5.37	2.56	.09	.09	**.90**	.82
6	3.27	1.74	.09	.11	**.89**	.81
8	3.46	1.43	.08	.06	**.75**	.58
10	2.93	1.38	.06	.10	**.62**	.39

Note. Boldface indicates highest factor loadings. Description of items found in Appendix A. Factor 1 = External Reasons (Out of Student's Control); Factor 2 = Student Effort; Factor 3 = Support From Others; h^2 = communality.

Table 9.7

Table X

Principal-Components Analysis With Varimax Rotation and

Coefficient Alphas

Coefficient alphas for items loading highly together on each factor are provided.

Item	Factor loadings
Factor 1: External Reasons (Out of Student's Control)	
(α = .70)	
3	.89
14	.86
9	-.86
5	.79
11	.79
1	.77
Factor 2: Student Effort (α = .62)	
7	.86
15	.79
2	.77
13	.75
4	.69
Factor 3: Support From Others (α = .54)	
12	.90
6	.89
8	.75
10	.62

Note. Item descriptions can be found in Appendix A.

Example 3

In this example, the SFQ is factor analyzed using an oblique (oblimin) rotation (Table 9.8). Because an oblimin rotation is used, the researcher should present the correlations among the factors as illustrated in Table 9.8.

This is the "Play It Safe" table for an obliquely rotated factor analysis solution.

Table 9.8

Table X

Summary of Factor Loadings for Oblimin Three-Factor Solution for the Student Failure Questionnaire

	Factor loading		
Item	1	2	3
1. Student is excessively absent from school.	.65	.85	.32
2. Student does extra work on his or her own.	.34	.90	.54
3. Education is taught in a second language.	.88	.41	.03
4. Student asks questions about material.	.34	.55	.91
5. Student lacks adequate diet.	.99	.34	.53
6. Student lacks support from school.	.45	-.02	.78
7. Student does not do homework.	.57	.85	.44
8. Student lacks support from teacher.	.24	.43	.91
9. Student has attention deficit disorder.	.89	.68	.32
10. Student lacks support from other students.	.43	.42	.72
11. Student appears to be of superior intellect.	.92	.31	.02
12. Student lacks parental support.	.42	.38	.68
13. Student makes extra effort to participate in class.	.34	.55	.24
15. Student enjoys doing schoolwork.	.34	.63	.03
14. Student has learning disability.	.45	.46	.34
Factor correlations			
Factor 1	--		
Factor 2	.39	--	
Factor 3	.26	.31	--

Table 9.11

> For each sample, items have been ordered according to their factor loadings.

Table X

Factor Scores for Student Failure Questionnaire Items for

Teacher and Parent Raters

Teachers		Parents	
Item	Factor loading	Item	Factor loading
Factor 1: External Reasons (Out of Student's Control)			
3	.89	5	.70
14	.86	9	-.65
9	-.86	3	.62
5	.79	11	.62
11	.79	14	.58
1	.77	1	.59
Factor 2: Student Effort			
7	.86	2	.83
15	.79	4	.73
2	.77	7	.71
13	.75	15	.69
4	.69	13	.56
Factor 3: Support From Others			
12	.90	8	.72
6	.89	6	.69
8	.75	10	.64
10	.62	12	.52

Note. Item descriptions can be found in Appendix A.

CHAPTER 10

Frequency and Demographic Data

What Is It?

Frequency data are used to summarize the number of cases or instances of a particular characteristic or variable. Demographic information provides a summary of participant characteristics (e.g., age, occupation).

What Tables Are Used?

Frequency data are included in a table only if they are particularly important in a study, for example, if the dependent measure is frequencies of particular behaviors. Demographic information usually is presented in a table if the participants are a special population (e.g., a clinical sample, an animal sample, or an organizational sample). Often, the data that are presented in a demographic table are frequencies (e.g., number of participants in each age group). It should be noted that demographic information can be displayed in either a frequency table or a table of means and standard deviations or in a combination of the two.

"Play It Safe" Table

For demographic data, the "safe" choice is Table 10.1 because it is more comprehensive. For a frequency table, the "safe" choice is Table 10.3.

Example

A group of researchers are interested in maternity leave in major corporations. Specifically, the researchers want to determine women's attitudes toward their corporations' maternity leave policies. The researchers surveyed 1,022 women who had taken maternity leave from corporations in major North American cities. The survey used asked the women questions about many aspects of their maternity leaves and corporation policies, including length, number of leaves per employee, and benefits during leave. In summarizing their research, the researchers created a demographics table of the characteristics of the women surveyed (Table 10.1 or 10.2) and a frequency table to summarize their results (Table 10.3).

Exhibit 10.1

Variables
1. Maternity leave policy
2. Attitude toward maternity leave policy

Table 10.1

Table X

Demographic Characteristics of Participants (N = 1,022)

Characteristic	n	%
Age at time of survey (years)		
20-29	244	24
30-39	534	52
40-49	132	13
50-59	112	11
Age at time of maternity leave (years)		
20-29	122	12
30-39	834	82
40-49	66	6
Highest education level completed		
High school	245	24
Undergraduate school	441	43
Graduate school	133	13
Professional school	203	20
Annual income ($)		
0-14,999	129	13
15,000-29,999	201	20
30,000-44,999	309	30
45,000-59,999	211	21
60,000-74,999	109	11
75,000-89,999	42	4
90,000-104,999	19	2
105,000+	2	<1
Length of leave (weeks)		
0-3	110	11
4-6	243	24

(Table X continues)

(Table X continued)

Characteristic	n	%
7-9	286	28
10-12	198	19
13-15	155	15
16-18	24	2
19-21	4	<1
22+	2	<1
Average pay received during leave (%)		
0-24	134	13
25-49	300	29
50-74	278	28
75-99	234	23
100+	76	7
Number of leaves taken		
1	502	49
2	322	31
3	159	16
4	39	4

Table 10.2

Table X

Participant Characteristics (N = 1,022)

Characteristic	M	SD
Age at time of survey (years)	37.12	8.21
Age at time of maternity leave (years)	32.33	4.13
Years of education	15.34	3.04
Annual income ($)	38,723	15,201
Length of leave (weeks)	7.65	3.95
Average pay received during leave (%)	43	20
Number of leaves taken	1.66	0.71

This is the "Play It Safe" table for frequency information.

This table presents the frequency of responses for one of the survey questions.

Table 10.3

Table X

Responses to Survey Question "What Would You Change About Your Employer's Maternity Leave Policy? (Choose One)"

Response	n	%
Nothing	33	3
Length of paid leave	145	14
Hiring of replacement for leave	22	2
Salary during leave	567	56
Stigma associated with leave	99	10
Benefits during leave	114	11
Layoff protection during leave	42	4

Note. N = 1,022.

CHAPTER 11

Logistic Regression

What Is It?

Logistic regression is a variant of multiple regression; the procedure assesses the relation between one criterion (dependent) variable and several predictor (independent) variables. In logistic regression the criterion variable is categorical and the predictor variables usually include both categorical and continuous variables. Logistic regression analysis allows the researcher to estimate the odds of an event (one level of the dependent variable) occurring on the basis of the values for the predictor variables.

What Tables Are Used?

There are three tables that are most commonly used to report the results of a logistic regression analysis: (a) a table of means and frequencies (Table 11.1), (b) a table of intercorrelations for predictor and criterion variables (Table 11.2), and (c) a logistic regression summary table (Table 11.3 or 11.4).

"Play It Safe" Table

The "Play It Safe" table for logistic regression is Table 11.3. Note that to be "safe," a table of means and frequencies (Table 11.1) and a table of intercorrelations (Table 11.2) should be included.

Example

A group of researchers are conducting a study in which they are investigating factors related to infantile amnesia. *Infantile amnesia* refers to the lack of very early childhood memories demonstrated by most adults. The researchers hypothesized that infantile amnesia should be less severe for adults who, as young children, spent a considerable amount of time talking and interacting with adults. The participants are 140 college students. Half of the college students can recall at least one event that happened before they were 3.5 years old, and the other half of the students have no memories of before they were 3.5 years old. Thus, the first group of students have an early childhood memory, whereas the second group of students do not. These are the two levels of the criterion (dependent) variable. The predictor (independent) variables are (a) whether the participant was an only child, (b) frequency of visits with grandparents, (c) IQ, (d) verbal fluency, and (e) working memory span.

Exhibit 11.1

Independent Variables
  1. Whether only child
  2. Frequency of visits with grandparents
  3. IQ
  4. Verbal fluency
  5. Working memory span

Dependent Variable
  1. Early childhood memory (presence or absence of)

Table 11.1

> For other examples of tables of means and standard deviations, frequencies, and a priori contrasts, see chapters 13, 10, and 18, respectively.

Table X

Mean Values or Frequencies for Predictor Variables as a Function of Early

Childhood Memory

Variable	Memory before 3.5 years old ($\underline{n}$ = 70)	No memory before 3.5 years old ($\underline{n}$ = 70)	$\chi^2(1)$ or $\underline{t}(138)$
Only child (%)	53	29	4.83*
Grandparent visits	8.17	5.56	3.12***
IQ	111.20	102.01	4.53***
Verbal fluency	47.13	50.54	−2.84*
Working memory span	8.03	7.43	1.64

Note. Chi-square test used for child variable; $\underline{t}$ test used for all other

variables.

*$\underline{p}$ < .05. ***$\underline{p}$ < .001.

Table 11.2

Table X

Intercorrelations for Early Childhood Memory and Predictor Variables

Measure	1	2	3	4	5	6
1. Early childhood memory	--					
2. Only child	.25**	--				
3. Grandparent visits	.26**	.03	--			
4. IQ	.36**	.01	.39**	--		
5. Verbal fluency	−.23**	−.09	−.22**	−.05	--	
6. Working memory span	.14	.06	.16	.18*	−.06	--

Note. Early childhood memory coded as 1 = no memory before age 3.5,

2 = memory before age 3.5. Only child coded as 1 = not only child, 2 = only

child.

*$\underline{p}$ < .05. **$\underline{p}$ < .01.

This is the "Play It Safe" table for a logistic regression summary.

Table 11.3

The 95% confidence interval for the odds ratio could be included in an additional column.

Table X

Summary of Logistic Regression Analysis Predicting Early Childhood Memory

Variable	B	SE	Odds ratio	Wald statistic
Only child	1.13	0.40	0.32	7.83**
Grandparent visits	0.06	0.04	0.94	1.67
IQ	0.07	0.02	0.94	12.66**
Verbal fluency	-0.07	0.03	1.07	5.44*
Working memory span	0.06	0.09	0.94	0.42

*p < .05. **p < .01.

This is the most common format for a logistic regression summary table.

Table 11.4

Table X

Logistic Regression Predicting Early Childhood Memory

Predictor	β	SE	Odds ratio
Only child	0.37**	0.40	0.32
Grandparent visits	0.01	0.04	0.94
IQ	0.33**	0.02	0.94
Verbal fluency	-0.23*	0.03	1.07
Working memory span	0.02	0.09	0.94

*p < .05. **p < .01.

CHAPTER 12

Log-Linear Analysis

What Is It?

Log-linear analysis is used to examine the relation between two or more categorical variables to determine the best model that will account for the observed frequencies. In log-linear analysis, all of the categorical variables are considered to be independent variables. There are several ways to conduct log-linear analysis (e.g., hierarchical or nonhierarchical), but each requires that results be presented in a similar manner. A hierarchical log-linear analysis using backward elimination is used as the example in this chapter.

What Tables Are Used?

There are three tables that are used to present the results of log-linear analysis: (a) a table of the observed frequencies for the variables (see Tables 12.1 and 12.4), (b) a table of the log-linear parameters and the goodness-of-fit tests (see Table 12.2), and (c) a table of the goodness-of-fit index for each step in the analysis (see Table 12.3). Table 12.2 is not necessary if the final model consists of only a few main effects. In that case, the results can be reported in the text of the Results section rather than in a table.

"Play It Safe" Table

The "safe" choices for log-linear analysis include Table 12.1 (observed frequencies for all of the variables), Table 12.2 (log-linear parameters, z values, and goodness-of-fit tests) or Table 12.3 (goodness-of-fit index for each step), and Table 12.4 (observed frequencies for significant interactions). If many variables are included in the analysis, Table 12.1 can be omitted, because a table of observed frequencies for all of the variables would be too complex.

Example

A physical education instructor wishes to determine the relations among gender, sport, and major. A total of 503 college students completed the survey. The instructor categorized sport activities into three categories: (a) individual sport (e.g., jogging); (b) team sport (e.g., hockey); and (c) a cross between individual sport and team sport, which was labeled *both* (e.g., swim team). The instructor grouped the students' course majors into two broad categories: arts or science. Thus, there are three variables in this analysis: gender, sport, and major.

The observed cell frequencies are presented in Table 12.1. The results of the log-linear analysis conducted on the data are presented in Tables 12.2 and 12.3. The observed cell frequencies for the interaction included in the final model are given in Table 12.4.

Exhibit 12.1

Variables
  1. Gender (men or women)
  2. Sport (individual, team, or both)
  3. Major (arts or science)

Table 12.1

Table X

Observed Frequencies and Percentages for Gender, Sport,

and Major

Sport	Women	Men
	Arts major	
Team	64 (13%)	47 (9%)
Individual	30 (6%)	14 (3%)
Both	14 (3%)	14 (3%)
	Science major	
Team	30 (6%)	14 (3%)
Individual	91 (18%)	45 (9%)
Both	93 (18%)	47 (9%)

This table is not always used, particularly when there are many variables included in the log-linear analysis, making the table difficult to read. In such a case, the researcher may wish to present only the observed frequencies for effects that are included in the model (e.g., significant interactions).

This table, along with Table 12.2 or 12.3 and Table 12.4, is the "Play It Safe" table for presenting the results of log-linear analysis.

Table 12.2

Table X

Log-Linear Parameter Estimates, Values, and

Goodness-of-Fit Index for Gender, Sport, and Major

Effect	Coefficient	z
Sport × Major	.764	9.90*
	−.281	−3.66*
Gender	.264	4.76*
Sport	.005	0.07
	.082	1.07
Major	−.280	−5.04*

Note. $G^2(5, N = 503) = 5.73$, $p > .05$.

*$p < .05$.

Note that only significant effects are presented in this table. In other words, only the effects that describe the final model, as determined by the log-linear analysis, are included.

Standard errors could be presented instead of z values.

> The researcher may wish to present the goodness-of-fit index for each step of the analysis as shown in this table.

Table 12.3

Table X

Summary of Hierarchical Deletion Steps Involved in Arriving at Final Model

Step	Model	df	G^2	p	Term deleted	Δdf	ΔG^2	Δp
1	(GrS)(GrM)(SM)	2	1.93	.381				
					GrS	2	0.70	.705
2	(GrM)(SM)	4	2.63	.621				
					GrM	1	3.10	.078
3	(SM)(Gr)	5	5.73	.333				
					Gr	1	40.06	.000

Note. Gr = gender; S = sport; M = major.

> Note that the three-way interaction (Gender × Sport × Major) is not presented in this table because it is the saturated model and therefore has zero degrees of freedom and a G^2 also equal to zero. Also, terms listed imply that lower order terms (i.e., main effects in this example) are included. The last step presented is the one in which Δp is significant.

Table 12.4

Table X

Cross-Tabulation of Observed Frequencies and Percentages

for Sport × Major Interaction

Sport	Major	
	Arts	Science
Team	111 (22%)	44 (9%)
Individual	44 (9%)	136 (27%)
Both	28 (6%)	140 (28%)

> A frequency table is useful for displaying any interactions that may be in the final model.

CHAPTER 13

Means

What Is It?

A *mean* is a measure of central tendency. It is also called the *average*.

What Tables Are Used?

One table is used, and it includes the means and standard deviations (see Tables 13.1, 13.2, 13.3, and 13.7). There are numerous ways in which this information can be presented. Sometimes the number of participants is included (see Tables 13.4, 13.5, and 13.6). If there is little space, such as when there are many means, they may be presented without their standard deviations (see Table 13.4). It is important to consider that means often are presented in graphs rather than tables (note that no graphical examples are included in this chapter) or in tables in combination with other results such as coefficient alphas, intercorrelations, percentage of spoiled trials removed from analyses, and so on. Example 1 presents the results of the effects of stress on a single dependent variable, whereas Example 2 presents the results of the effects of stress on two dependent variables.

"Play It Safe" Table

The most comprehensive table includes the means, standard deviations, and number of participants. If the number of participants in each cell is the same, it is not necessary to include this information (see Tables 13.1, 13.2, and 13.3). If the number of participants varies from cell to cell, then this information would be useful (see Tables 13.5 and 13.6).

Example 1

Three species of male rats (Species X, Y, and Z), all of the same weight and age, were randomly assigned to one of two conditions. The two conditions represent two different stress environments. *Stress* is operationally defined here as exposure to 85-dB heavy metal music. Rats in the stressful environment were exposed to the heavy metal music 6 hr per day for 30 days. The rats in the control group were maintained in their usual environment (i.e., no exposure to any kind of music). The goals of this study were to examine the effects of a stressful environment on the amount of food eaten and to determine whether different species of rats are differentially affected by the stressful situation examined. The amount of food eaten by the different species of rats was recorded for 30 days.

There are two independent variables: species and stress environment. The independent variable, species, has three levels (Species X, Species Y, and Species Z). The independent variable, stress environment, has two levels (high stress [85-dB] heavy metal music and low stress [no music]). The dependent variable is the amount of rat chow eaten.

Exhibit 13.1

Independent Variables
1. Rat species (X, Y, or Z)
2. Stress environment (high or low stress)

Dependent Variable
1. Amount of rat chow eaten (in grams)

Researchers often will present descriptive statistics before presenting their analyses. Tables 13.1 through 13.6 provide various alternatives for presenting means and standard deviations. In particular, Tables 13.4, 13.5, and 13.6 include the number of participants.

Table 13.1

Table X

Mean Amount of Rat Chow Eaten (in Grams) and Standard

Deviations for Three Species of Rats and Two Stress

Conditions

	High stress		Low stress	
Species	M	SD	M	SD
X	702.68	9.21	713.54	10.62
Y	721.39	14.76	724.76	15.98
Z	717.25	16.19	729.14	14.76

Table 13.2

Table X

Condition Means (in Grams) for Species X, Y, and Z in Two

Stress Conditions

Species	High stress	Low stress
X	702.68 ± 9.21	713.54 ± 10.62
Y	721.39 ± 14.76	724.76 ± 15.98
Z	717.25 ± 16.19	729.14 ± 14.76

Note. Values are M ± SD.

Table 13.3

Table X

Mean Amount of Rat Chow Eaten (in Grams) and

Standard Deviations for Three Species of Rats Under

High- or Low-Stress Conditions

Species	High stress	Low stress
X		
M	702.68	713.54
SD	9.21	10.62
Y		
M	721.39	724.76
SD	14.76	15.98
Z		
M	717.25	729.14
SD	16.19	14.76

> This is a "Play It Safe" table of means and standard deviations when the number of participants in each cell is equal.

Table 13.4

Table X

Mean Amount of Rat Chow Eaten (in Grams) by Three Species

of Rats in High-Stress (85-dB Heavy Metal Music) and

Low-Stress (No-Music) Conditions

Species	M Total (N = 300)	M High stress (n = 150)	M Low stress (n = 150)
X	708.11	702.68	713.54
Y	723.08	721.39	724.76
Z	723.20	717.25	729.14

> It is recommended that standard deviations be included. However, if there are many means, standard deviations may be eliminated to save space.

Table 13.5

This is a "Play It Safe" table of means and standard deviations when the number of participants in each cell varies.

Table X

Mean Grams of Rat Chow Eaten by Three Species of
Rats in High- and Low-Stress Conditions

Species	High stress	Low stress
X		
M	702.68	713.54
SD	9.21	10.62
n	45	49
Y		
M	721.39	724.76
SD	14.76	15.98
n	47	54
Z		
M	717.25	729.14
SD	16.19	14.76
n	58	47

This is a "Play It Safe" table of means and standard deviations when the number of participants in each cell varies.

Table 13.6

Table X

Mean Amount of Rat Chow Eaten (in Grams) by Three Species of
Rats in Two Stress Conditions

Species	High stress			Low stress		
	M	SD	n	M	SD	n
X	702.68	9.21	45	713.54	10.62	49
Y	721.39	14.76	47	724.76	15.98	54
Z	717.25	16.19	58	729.14	14.76	47

Example 2

This is the same study as described in Example 1, but the effects of stress and species are examined on two dependent variables: (a) amount of rat chow eaten and (b) amount of pacing.

Exhibit 13.2

Independent Variables
1. Rat species (X, Y, or Z)
2 Stress environment (high or low stress)

Dependent Variables
1. Amount of rat chow eaten (in grams)
2. Amount of pacing (in centimeters per minute)

Table 13.7

Table X

Amount of Rat Chow Eaten (in Grams) and Amount of Pacing (Centimeters per Minute) by Three Species of Rats in Two Stress Conditions

Species	High stress M	High stress SD	Low stress M	Low stress SD
	Amount of rat chow eaten			
X	702.68	9.21	713.54	10.62
Y	721.39	14.76	724.76	15.98
Z	717.25	16.19	729.14	14.76
	Amount of pacing			
X	100.25	21.34	65.23	8.56
Y	135.21	28.12	87.34	10.01
Z	98.87	17.03	45.32	9.45

CHAPTER 14

Meta-Analysis

What Is It?

Meta-analysis is a statistical technique that is used to combine the results of several independent studies. Meta-analysis typically is used to describe the population distribution of an effect size, a correlation, or a reliability coefficient.

What Tables Are Used?

Most researchers use two tables to present the results of a meta-analysis: (a) a table that summarizes the characteristics of the studies included in the meta-analysis and (b) a table of the results of the meta-analysis. Sometimes, however, researchers present only the first type of table. This is usually the case if the researcher is interested only in the overall effect size, reliability, or correlation of the results in the meta-analysis and is not interested in how the effect size might differ depending on the studies' various moderating characteristics. These researchers will present one table containing the characteristics of the studies and usually will provide the overall effect size, reliability, or correlation in the text (see Example 1 and Tables 14.1 and 14.2). Researchers who present both types of tables are interested in whether the effect size, reliability, or correlation differs in different categories of studies (i.e., how the effect size is affected by various moderating variables) and use the second type of table to present those findings (see Example 2 and Tables 14.3, 14.4, 14.5 and 14.6).

"Play It Safe" Table

If a researcher is interested only in the overall effect size, reliability, or correlation produced by the meta-analysis and has not made any hypotheses about how these might vary depending on different characteristics of the previous studies, then Table 14.2 is the "safe" choice. If, however, the researcher has hypotheses about moderating variables or different categories of studies, then both Tables 14.2 and 14.5 constitute the "Play It Safe" choice.

Example 1

In this example, the researchers are interested in attitude change. Specifically, they are interested in determining how attitudes are affected by endorsement from an attractive versus a nonattractive speaker. There have been many previous studies on this aspect of attitude change, and these researchers want to perform a meta-analysis on those studies to establish whether there is a significant overall effect for attitude change depending on endorser attractiveness. After searching different research databases (e.g., PsycINFO), the researchers chose 15 studies that satisfied their criteria for inclusion in the meta-analysis. (Note: All of the studies used in this example are fictitious.)

For this example, the dependent variable is attitude change, which has been assessed in many different ways in the studies examined by the researchers. The independent variable is the attractiveness of the speaker or endorser.

Exhibit 14.1

Independent Variable
1. Endorser attractiveness (attractive vs. nonattractive)

Dependent Variable
1. Attitude change

> This type of summary table often is omitted; instead, the studies included in the meta-analysis are identified with asterisks in the reference list.

Table 14.1

Table X

Summary of Studies Included in Meta-Analysis on Attitude Change and Endorser Attractiveness

Study	n	d	SD
Bigness & Holmes (1982)	56	0.20	0.21
Brank & Jones (1991)	145	1.32	0.98
Calvert & Snell (1987, Experiment 2)	69	0.24	0.22
Cruikshank et al. (1996)	77	1.85	1.12
Czewski (1988)	88	1.22	0.88
Dunkley, Rogers, & Graham (1992)	102	0.17	0.09
Goodchild & Harding (1991a)	100	1.70	1.45
Goodchild & Harding (1991b, Experiment 3)	56	1.44	1.41
Goodchild & Kacinik (1993)	50	2.08	1.55
Harrison (1989, Experiment 1)	156	0.37	0.23
Harrison et al. (1986)	98	0.46	0.44
Jenkins & Harrison (1990)	28	0.24	0.21
Melville & Harrison (1985)	58	1.28	0.88
Smith & Crane (1989)	122	0.91	0.55
Smith & Smith (1988, Experiment 2)	78	0.94	0.66

> In this sample table, d is used to express effect size, but there are other values that might be presented instead (e.g., *ES*, *Zr*), or the researcher could present reliability or correlation coefficients rather than effect size.

Because the outcome measures varied among the studies included in this meta-analysis, it could be useful to list these different measures in the summary table, as illustrated in this table.

This is the "Play It Safe" version of a summary table for a meta-analysis.

Table 14.2

Table X

Summary of Studies Included in Meta-Analysis on Attitude Change and Endorser Attractiveness

Study	Indicator of attitude change	n	d	95% CI Lower limit	95% CI Upper limit
Bigness & Holmes (1982)	Opinion on affirmative action	56	0.20	0.10	0.30
Brank & Jones (1991)	Opinion on standardized tests	145	1.32	1.12	1.42
Calvert & Snell (1987, Experiment 2)	Opinion on abortion	69	0.24	0.11	0.37
Cruikshank et al.(1996)	Opinion on spanking children	77	1.85	1.42	2.28
Czewski (1988)	Opinion on universal health care	88	1.22	0.99	1.45
Dunkley et al. (1992)	Opinion on euthanasia	102	0.17	0.09	0.24
Goodchild & Harding (1991a)	Opinion on capital punishment	100	1.70	1.45	1.95
Goodchild & Harding (1991b, Experiment 3)	Opinion on capital punishment	56	1.44	1.23	1.63

(Table X continues)

(Table X continued)

Study	Indicator of attitude change	n	d	95% CI Lower limit	Upper limit
Goodchild & Kacinik (1993)	Opinion on capital punishment	50	2.08	1.78	2.38
Harrison (1989, Experiment 1)	Opinion on Title IX	156	0.37	0.27	0.47
Harrison et al. (1986)	Opinion on affirmative action	98	0.46	0.31	0.61
Jenkins & Harrison (1990)	Opinion on abortion	28	0.24	0.15	0.33
Melville & Harrison (1985)	Opinion on abortion	58	1.28	1.12	1.44
Smith & Crane (1989)	Opinion on abortion	122	0.91	0.71	1.11
Smith & Smith (1988, Experiment 2)	Opinion on abortion	78	0.94	0.69	1.19

Note. CI = confidence interval.

Example 2

In this study, the researchers are interested in attitude change and endorser attractiveness, but they also are interested in how this relationship is moderated by several variables. They hypothesize that effect size might be moderated by (a) the sex of the endorser relative to the sex of the participant (endorsers who are the opposite sex to the participants will be most effective), (b) the age of the endorser relative to the age of the participant (endorsers who are older than the participants will be more effective), and (c) the mode of presentation for the endorsement (endorsements in person will be more effective than endorsements on videotape).

For this example, the researchers would include a summary table of the individual studies included in the meta-analysis (such as Table 14.1 or 14.2), but they would also include a table of the meta-analysis results (such as Tables 14.3 through 14.7).

Table 14.3

Table X

Summary Statistics for Total Sample in Meta-Analysis and for Three
Moderating Variables

Category	n	k	d	SD
Total sample	1,273	15	0.92*	0.23
Sex of target vs. endorser				
Opposite	240	4	0.96*	0.21
Same	362	8	0.25	0.15
Age of target vs. endorser				
Younger	244	3	0.83*	0.15
Same	204	4	0.36*	0.18
Older	455	6	1.44*	0.21
Presentation mode				
Videotape	256	5	0.40*	0.21
In person	828	10	1.06*	0.21

*p < .05.

Standard deviations for effect size are not always included in these tables and are sometimes expressed in different ways (i.e., σ_δ).

Table 14.4

Table X

Summary Statistics for Effects of Sex, Age, and Presentation Mode in

Meta-Analysis

| | | | 95% CI | | |
| | | | Lower | Upper | |
Category	$\underline{k}$	$\underline{d}$	limit	limit	$\underline{Q}$
Sex of target vs. endorser					
Opposite	4	0.96*	0.80	1.12	5.36
Same	8	0.25	0.12	0.38	10.40*
Age of target vs. endorser					
Younger	3	0.83*	0.60	1.06	3.20
Same	4	0.36*	0.22	0.40	9.64*
Older	6	1.44*	1.22	1.66	2.11
Presentation mode					
Videotape	5	0.40*	0.30	0.50	4.91
In person	10	1.06*	0.90	1.22	8.80*

Note. CI = confidence interval; $\underline{Q}$ = test of homogeneity.

*$\underline{p}$ < .05.

Table 14.5

Table X

Mean Effect Sizes for Various Moderators

Category	n	k	d	95% CI Lower limit	95% CI Upper limit	r	Q
Sex of target vs. endorser							
Opposite	240	4	0.96*	0.80	1.12	.49	5.36
Same	362	8	0.25	0.12	0.38	.13	10.40*
Age of target vs. endorser							
Younger	244	3	0.83*	0.60	1.06	.35	3.20
Same	204	4	0.36*	0.22	0.40	.18	9.64*
Older	455	6	1.44*	1.22	1.66	.71	2.11
Presentation mode							
Video	256	5	0.40*	0.30	0.50	.21	4.91
In person	828	10	1.06*	0.90	1.22	.51	8.80*

Note. CI = confidence interval; Q = test of homogeneity.

*p < .05.

Table 14.6

Table X

Summary of Effect Sizes for Moderating Variables of Sex, Age, and Presentation Mode

Category	k	d	r	Q
Sex of target vs. endorser				
Opposite	4	0.96*	.49	5.36
Same	8	0.25	.13	10.40*
Age of target vs. endorser				
Younger	3	0.83*	.35	3.20
Same	4	0.36*	.18	9.64*
Older	6	1.44*	.71	2.11
Presentation mode				
Videotape	5	0.40*	.21	4.91
In person	10	1.06*	.51	8.80*

Note. Q = test of homogeneity.

*p < .05.

This is the most common table for illustrating moderating effects in meta-analysis because these are the four statistics that are most commonly presented. However, the columns are not always in the order presented here.

> This is the most brief table for presenting meta-analysis results.

Table 14.7

Table X

Summary Statistics for Three Moderating Variables in Meta-Analysis

| | | 95% CI | |
| | | Lower | Upper |
Category	d	limit	limit
Sex of target vs. endorser			
Opposite	0.96*	0.80	1.12
Same	0.25	0.12	0.38
Age of target vs. endorser			
Younger	0.83*	0.60	1.06
Same	0.36*	0.22	0.40
Older	1.44*	1.22	1.66
Presentation mode			
Videotape	0.40*	0.30	0.50
In person	1.06*	0.90	1.22

Note. CI = confidence interval.

*p < .05.

Multiple Regression

What Is It?

Multiple regression is a statistical procedure that assesses the relation between one criterion (dependent) variable and several predictor (independent) variables. There are several types of multiple regression analyses (e.g., stepwise, forward, backward), which differ in the way independent variables are entered into the regression equation.

What Tables Are Used?

There is considerable variability in the way results from multiple regression analyses are presented. That is, there are many versions of multiple regression tables in the literature. This variability is based on the purpose of the particular multiple regression table. Some researchers consider more aspects of multiple regression analyses than others and hence provide more comprehensive tables. Typically, when reporting the results of multiple regression analyses, two tables are used: (a) a table of means, standard deviations, and correlations for predictor and dependent variables (Table 15.1) and (b) a multiple regression analysis summary table (Tables 15.2, 15.3, 15.4, 15.5, 15.6, or 15.7).

"Play It Safe" Table

The "Play It Safe" table for a standard multiple regression analysis is Table 15.2. The "Play It Safe" table for a hierarchical multiple regression analysis is Table 15.3. Note also that to be "safe," a table of means, standard deviations, and intercorrelations should be included (Table 15.1).

Example

This example is a study in which researchers are examining the effect of parent characteristics on children's early phonemic awareness. Phonemic awareness tends to predict reading success, and the researchers want to determine parent characteristics that are related to that early awareness. Their study includes 68 preschool children and their parents. The dependent (criterion) variable is phonemic awareness, as measured by a test the researchers call the Phonemic Awareness Measure (PAM). On the PAM, children are asked to choose, for example, which words start with the same sound. The independent (predictor) variables are the parent characteristics: education level, literacy, hours per week of reading to the child, hours per week of own reading, and socioeconomic status.

Exhibit 15.1

Independent Variables
1. Parental education level
2. Parental literacy
3. Hours per week of reading to child
4. Hours per week of parent's own reading
5. Socioeconomic status

Dependent Variable
1. Phonemic awareness (Phonemic Awareness Measure score)

There are many alternate formats for correlation tables and tables of means and standard deviations. See chapters 7 and 13, respectively.

Table 15.1

Table X

Means, Standard Deviations, and Intercorrelations for Children's Phonemic

Awareness and Parent Characteristics Predictor Variables

Variable	M	SD	1	2	3	4	5
Phonemic Awareness Measure	45.3	11.1	.66*	.59*	.55*	.46*	.32*
Predictor variable							
1. Parental education level	12.6	3.2	--	.88**	.86**	.91**	.79**
2. Parental literacy	9.2	2.1		--	.77**	.88**	.78**
3. Reading to child	2.2	0.9			--	.82**	.75**
4. Parents' own reading	3.4	1.3				--	.59**
5. Socioeconomic status	9.6	3.7					--

*p < .05. **p < .01.

Table 15.2

Table X

Regression Analysis Summary for Parent Variables Predicting

Children's Phonemic Awareness

Variable	B	SEB	β
Parental education level	.29	.08	.58*
Parental literacy	.19	.09	.13
Reading to child	.25	.07	.46*
Parents' own reading	.12	.09	.18
Socioeconomic status	.31	.14	.49*

Note. R^2 = .61 (N = 68, p < .01).

*p < .05.

This is the "Play It Safe" table for a standard multiple regression analysis.

The R^2 value could be presented in the text rather than in a table note.

Note that either the adjusted or unadjusted R^2 value may be presented.

Table 15.3

> This is the "Play It Safe" table for a hierarchical multiple regression analysis.

Table X

Hierarchical Regression Analysis Summary for Parent Variables

Predicting Children's Phonemic Awareness (N = 68)

Variable	B	SEB	β	R^2	ΔR^2
Step 1				.32*	
Parental education level	1.34	.34	.30*		
Step 2				.39*	.07
Reading to child	1.33	.47	.23*		
Step 3				.51*	.12
Parents' own reading	0.91	.33	.23*		
Step 4				.54*	.03
Parental literacy	0.51	.30	.13		
Step 5				.76*	.22*
Socioeconomic status	1.23	.09	.25		

*p < .05.

Table 15.4

Table X

Prediction by Parent Predictor Variables of Children's Phonemic

Awareness

Hierarchical step	Predictor variable	Total R^2	Incremental R^2
1	Parental education level	.32*	.32*
2	Reading to child	.39*	.07
3	Parents' own reading	.51*	.12
4	Parental literacy	.54*	.03
5	Socioeconomic status	.76*	.22*

*p < .05.

For this table, the analysis was changed (in terms of the variables included in each step) to illustrate inclusion of semipartial correlations.

Table 15.5

Table X

Hierarchical Regression Analysis Predicting Phonemic Awareness With Parent Variables

Step and predictor variable	R^2	ΔR^2	sr^2	β
Step 1	.32*	.32*		
Parental education level			.31*	.30*
Reading to child			.38*	.23*
Step 2	.42	.10		
Parents' own reading			.51*	.23*
Parental literacy			.53*	.13
Socioeconomic status			.55*	.25*

*p < .05.

.32 in ΔR^2 column is redundant; it reflects that nothing was changed.

Table 15.6

Table X

Summary of Hierarchical Multiple Regression Analysis With Children's Phonemic Awareness as Criterion

Step	Predictor variable	R^2	ΔR^2	ΔF
1	Parental education level	.32	.32	21.01*
2	Reading to child	.39	.07	2.10
3	Parents' own reading	.51	.12	1.18
4	Parental literacy	.54	.03	1.01
5	Socioeconomic status	.76	.22	15.78*

*p < .05.

The .32 in ΔR^2 column demonstrates that this is the first entry and that no variable has been removed.

This is the briefest version of a multiple regression table.

Table 15.7

Table X

Hierarchical Multiple Regression Analysis Relating Parent

Variables to Children's Phonemic Awareness

Step and predictor variable	β	ΔR^2
1. Parental education level	.30*	.32*
2. Reading to child	.23*	.07
3. Parents' own reading	.23*	.12
4. Parental literacy	.13	.03
5. Socioeconomic status	.25*	.22*

*p < .05.

CHAPTER 16

Multivariate Analysis of Covariance

What Is It?

The multivariate analysis of covariance (MANCOVA) is the extension of an analysis of covariance (ANCOVA) to the situation in which there are several dependent variables and one or more covariates.

What Tables Are Used?

There are three types of tables that are especially relevant when presenting data that have been analyzed using a MANCOVA: (a) a table of means or adjusted means and standard deviations as a function of the independent variables (see chapter 2 on ANCOVA), (b) a table of correlations among the dependent variables (see chapter 17 on multivariate analysis of variance [MANOVA]), and (c) the summary table of univariate ANOVAs (see chap. 3) and possibly MANOVAs (see chap. 17). However, researchers do not always include all three of these tables when presenting the results of a MANCOVA. Sometimes the multivariate results are presented only in the text. Additionally, in some cases, the tables described in (a) and (b) or (a) and (c) are combined into a single table.

"Play It Safe" Table

The "safe" choice is to present each of the three types of tables described.

CHAPTER 17

Multivariate Analysis of Variance

What Is It?

The multivariate analysis of variance (MANOVA) is an extension of the analysis of variance to the situation in which there is more than one dependent variable.

What Tables Are Used?

There are three tables that are especially relevant for a MANOVA. The three most commonly used tables are (a) a table of means and standard deviations for the dependent variables (Table 17.1), (b) a table of correlations among the dependent variables (Table 17.2), and (c) a multivariate and univariate analyses of variance (ANOVAs) summary table (Tables 17.3 or 17.4). The results reported depend on the nature of the data as well as the relevance of the tables to the hypotheses and the context in which the tables are being presented (e.g., journal article vs. thesis).

"Play It Safe" Table

The "Play It Safe" tables are Tables 17.1, 17.2, and 17.3. That is, to be comprehensive, all three tables should be included.

Example

The example used in this chapter is a reading study. The researchers are interested in sex differences in reading impairment. Nondyslexic and dyslexic boys and girls are tested on four measures of reading ability. Thus, this is a 2 (boys or girls) × 2 (dyslexic or nondyslexic reader) MANOVA with four dependent variables. The dependent variables are four measures of reading ability: the Famous Authors Test, a verbal IQ test, the Ingersoll Reading Test, and average speed on a nonword pronunciation task (nonword performance).

Exhibit 17.1

Independent Variables
1. Gender (boys vs. girls)
2. Reading impairment (dyslexic vs. nondyslexic)

Dependent Variables
1. Famous Authors Test
2. Verbal IQ
3. Ingersoll Reading Test
4. Nonword pronunciation speed (nonword performance)

Table 17.2, an example of a correlation table, is especially relevant if there are significant correlations among some or all of the dependent variables. In the reading study example in this chapter, the four dependent variables are, for the most part, measures of the same thing: reading ability. Thus, these variables are highly correlated. The results of the multivariate and univariate ANOVAs are presented together in Table 17.3, whereas Table 17.4 illustrates a somewhat different version of a MANOVA summary table.

Table 17.1

Table X

Mean Scores and Standard Deviations for Measures of Reading Ability as a Function of Gender and Reading Impairment

See chapter 13 for other examples of format for tables of means and standard deviations.

| | Reading ability measure | | | | | | | |
| | Famous Authors Test | | Verbal IQ | | Ingersoll Reading Test | | Nonword performance | |
Group	M	SD	M	SD	M	SD	M	SD
Boys								
Dyslexic	26	3.4	105	11.2	15	3.3	1,090	89.2
Nondyslexic	27	4.5	106	11.1	24	4.4	820	56.7
Girls								
Dyslexic	25	3.5	102	10.1	16	3.5	1,010	78.0
Nondyslexic	30	2.9	109	12.3	25	3.9	801	49.6

Note. Nonword performance = average speed on a nonword pronunciation task.

Table 17.2

Table X

Correlation Coefficients for Relations Among Four Measures of Reading Ability

See chapter 7 for other examples of format for correlation tables.

Measure	Verbal IQ	Ingersoll Reading Test	Nonword performance
Famous Authors Test	.91**	.81**	.76*
Verbal IQ	--	.92**	.87**
Ingersoll Reading Test	--	--	.79*

Note. Nonword performance = average speed on a nonword pronunciation task.

*$p < .05$. **$p < .01$.

Table 17.3

Table X

Multivariate and Univariate Analyses of Variance for Reading Measures

				Univariate		
	Multivariate		Famous Authors	Ingersoll Verbal	Ingersoll Reading	Nonword
Source	df	F[a]	Test[b]	IQ[b]	Test[b]	performance[b]
Gender (G)	1	7.22**	9.19**	10.21**	2.12	0.03
Reading						
impairment (R)	1	23.07***	29.66***	25.44***	9.23***	6.41***
G × R	1	1.67	5.42***	1.11	2.41	2.31
MSE			2.14	0.64	0.92	0.99

Note. Multivariate F ratios were generated from Pillai's statistic. Nonword performance = average speed on a nonword pronunciation task.

[a]Multivariate df = 4, 119. [b]Univariate df = 1, 120.

**p < .01. ***p < .001.

> This is a somewhat different version of a MANOVA summary table than Table 17.3.

Table 17.4

Table X

Multivariate and Univariate Analyses of Variance F Ratios for Gender × Reading

Impairment Effects for Reading Measures

		ANOVA			
		Famous	Verbal	Ingersoll	Nonword
	MANOVA	Authors Test	IQ	Reading Test	performance
Variable	$F(4, 119)$	$F(1, 120)$	$F(1, 120)$	$F(1, 120)$	$F(1, 120)$
Gender (G)	7.22*	9.19**	10.21**	2.12	0.03
Reading					
impairment (R)	23.07*	29.66***	25.44***	9.23***	6.41***
G × R	1.67	5.42***	1.11	2.41	2.31

Note. F ratios are Wilks's approximation of Fs. MANOVA = multivariate analysis of variance; ANOVA = univariate analysis of variance; nonword performance = average speed on a nonword pronunciation task.

**$p < .01$. ***$p < .001$.

CHAPTER 18

Post Hoc and A Priori Tests of Means

What Is It?

Post hoc and a priori analyses are used to compare specific group means in studies whose nominal independent variables have more than two levels.

What Tables Are Used?

Often, the results of post hoc or a priori analyses are not presented in a table. If several of these analyses are conducted, however, then a table can be a useful way of summarizing the results. Generally, significant post hoc or a priori analyses are indicated in a table of means and standard deviations (Table 18.1 or 18.2).

"Play It Safe" Table

The "safe" choice for a post hoc or a priori table is Table 18.1. Although Tables 18.1 and 18.2 are equally comprehensive, Table 18.1 is the more conventional format for presenting post hoc or a priori analysis results.

Example

In this study, the researchers are interested in the creativity of severely depressed versus mildly depressed versus nondepressed individuals. A group of 60 severely depressed individuals is recruited from a psychiatric hospital. A group of 60 mildly depressed individuals is also recruited through the hospital. A third group of 60 individuals who have never been depressed is recruited using newspaper advertisements. All three groups of participants are assessed for creativity using four different measures: the Franklin Creativity Test, a rating of creative accomplishments, a creative writing exercise, and a peer rating of creativity. The researchers are interested in whether the three groups of participants differ significantly on any of these measures. In this study, the independent variable is participant group (severely depressed vs. mildly depressed vs. nondepressed). The dependent variables are the four measures of creativity.

Exhibit 18.1

Independent Variable
1. Participant group (severely depressed vs. mildly depressed vs. nondepressed)

Dependent Variables
1. Franklin Creativity Test
2. Creative accomplishments
3. Creative writing exercise
4. Peer rating

Table 18.1

The *n* for each participant group could be presented under each group name, particularly if group sizes vary.

Table X

Mean Scores on Four Measures of Creativity as a Function of Participant Group

Creativity measure	Participant group					
	Severely depressed		Mildly depressed		Nondepressed	
	M	SD	M	SD	M	SD
Franklin Creativity Test	12.8$_a$	6.7	10.2$_b$	5.8	6.3$_{a,b}$	4.4
Accomplishments	21.2$_a$	11.2	23.1$_b$	12.2	13.1$_{a,b}$	7.7
Writing exercise	8.7$_a$	3.3	6.7$_a$	2.9	5.5$_a$	3.3
Peer rating	9.1	3.4	9.0	4.1	8.0	3.1

Note. Means in a row sharing subscripts are significantly different. For all measures, higher means indicate higher creativity scores.

Different sets of subscripts can be used for each row in the table (i.e., $_{a,b}$ for row 1, $_{c,d}$ for row 2, etc).

This is the "Play It Safe" table for a post hoc or a priori analysis.

When using subscripts to denote significant contrasts, it is important to note that means in a row with the *same* subscript (i.e., $_a$) have been found to be significantly different.

Table 18.2

Table X

Creativity in Individuals With Severe Depression, With Mild Depression, and Who Are Nondepressed

Creativity measure	Severely depressed (1)		Mildly depressed (2)		Nondepressed (3)		Post hoc
	M	SD	M	SD	M	SD	
Franklin							
Creativity Test	12.8	6.7	10.2	5.8	6.3	4.4	3 < 1, 2
Accomplishments	21.2	11.2	23.1	12.2	13.1	7.7	3 < 1, 2
Writing exercise	8.7	3.3	6.7	2.9	5.5	3.3	3 < 2 < 1
Peer rating	9.1	3.4	9.0	4.1	8.0	3.1	3 = 2 = 1

Note. The numbers in parentheses in column heads refer to the numbers used for illustrating significant differences in the last column titled "Post hoc."

CHAPTER 19

Structural Equation Modeling

Confirmatory Factor Analysis: What Is It?

The purpose of confirmatory factor analysis is to test the goodness of fit of one or more hypothesized factor models of a measure.

What Tables Are Used?

Two types of tables are used: (a) a summary table of the fit indices for each hypothesized model (see Tables 19.1 and 19.2) and (b) a table of the factor loadings of the best-fit model (see Table 19.3). If only one or two models are being tested, then the fit indices may be presented within the text rather than in a table.

Other information such as the means and standard deviations of the items of the measure, coefficient alphas for the scales, and interscale correlations also could be presented. See chapter 13 for examples of tables of means and standard deviations and chapter 7 for correlation tables.

Example 1 presents the results of a confirmatory factor analysis for a single sample. Example 2 presents the results for two samples.

"Play It Safe" Table

There are two comprehensive tables: (a) the summary table listing the goodness-of-fit indices and (b) the table of the factor loadings (i.e., the standardized regression weights) for the best-fit model. If only one or two models are tested, the fit indices table is not needed; that information can be provided within the text. If more than two models have been tested, then a table of the fit indices should be presented. Tables 19.1, 19.3, 19.4, and 19.5 are "Play It Safe" tables.

Example 1

A new 15-item questionnaire to measure flexibility of thought was created. The questionnaire was given to 400 university students. The researchers wish to determine which of their three hypothesized factor structures best describe this new measure. The three models are as follows:

- Model 1—A single-factor model
- Model 2—A two-factor model consisting of (a) Ability to Adjust and (b) Openness to Ideas
- Model 3—A three-factor model consisting of (a) Ability to Compromise, (b) Openness to Ideas, and (c) Ability to Adjust

---

Exhibit 19.1

Variable
1. The 15 items of the Flexibility of Thought Questionnaire

Table 19.1

A summary table such as this is important, particularly if more than two models are being tested. If only one or two models are being tested, then the results can be given within the text.

Table X

Goodness-of-Fit Indices of Four Models (N = 400)

Model	df	χ^2	χ^2/df	AGFI	ECVI	RMSEA
Null	105	555.61***	5.29			
Single factor	106	526.24***	4.96	.76	.75	.061
Two factor	107	225.18***	2.10	.80	.79	.054
Three factor	108	88.42	0.82	.92	.90	.034

Note. AGFI = adjusted goodness-of-fit index; ECVI = expected cross-validation index; RMSEA = root-mean-square error of approximation.

***p < .001.

This table, along with a table of factor loadings, is the "Play It Safe" table for confirmatory factor analysis.

Other indices of fit, such as the normed fit index, nonnormed fit index, or normed comparative fit index, could be included in additional columns. Usually, three or more indices are included.

Table 19.2

In this table, the null model is not included.

Table X

Goodness-of-Fit Indicators of the Models for the Flexibility of Thought Questionnaire (N = 400)

Model	df	χ^2	χ^2/df	AGFI	ECVI	RMSEA
Single factor	106	526.24***	4.96	.76	.75	.061
Two factor	107	225.18***	2.10	.80	.79	.054
Three factor	108	88.42	0.82	.92	.90	.034

Note. AGFI = adjusted goodness-of-fit index; ECVI = expected cross-validation index; RMSEA = root-mean-square error of approximation.

***p < .001.

If the change in chi-square is to be presented, it can be included in a column immediately to the right of the column labeled χ^2/df.

Table 19.3

Standardized Solutions by Confirmatory Factor Analysis for the Three-Factor Model

	Factor		
Item	Ability to Compromise	Openness to Ideas	Ability to Adjust
10	.58		
5	.57		
14	.56		
9	.44		
3	.39		
12		.52	
2		.51	
8		.49	
1		.45	
6		.40	
13			.55
4			.50
15			.48
7			.43
11			.39

This table of factor loadings for the best-fit model, along with a table of fit indices, is the "Play It Safe" table for presenting the results of a confirmatory factor analysis.

For additional examples of these kinds of tables, see chapter 9.

Example 2

The Flexibility of Thought Questionnaire was given to two samples: (a) a group of students from a university located in a large city in Canada and (b) a group of students from an equally large city in the United States. The authors wish to determine how their three models (as identified in Example 1) fit each of these samples. Table 19.4 presents the goodness-of-fit indices and Table 19.5 presents the factor loadings for the best-fit model.

This table, along with a table of factor loadings, is the "Play It Safe" table for presenting the results of confirmatory factor analysis for two samples.

Table 19.4

Table X

Goodness-of-Fit Indicators of Models for the Flexibility of Thought

Questionnaire for Two Samples (N = 400)

Model	df	χ^2	χ^2/df	AGFI	ECVI	RMSEA
		Sample 1				
Single factor	106	526.24***	4.96	.76	.75	.061
Two factor	107	225.18***	2.10	.80	.79	.054
Three factor	108	88.42	0.82	.92	.90	.034
		Sample 2				
Single factor	106	502.37***	4.74	.78	.76	.060
Two factor	107	217.98***	2.04	.81	.80	.051
Three factor	108	79.35	0.73	.91	.91	.032

Note. AGFI = adjusted goodness-of-fit index; ECVI = expected cross-validation index; RMSEA = root-mean-square error of approximation.

***p < .001.

Other indices of fit could be included in additional columns.

Table 19.5

Table X

Factor Loadings of Flexibility of Thought Questionnaire Items for the Three-Factor Model for Two Samples

Item	Ability to Compromise		Openness to Ideas		Ability to Adjust	
	S1	S2	S1	S2	S1	S2
10	.58	.53				
5	.57	.56				
14	.56	.49				
9	.44	.50				
3	.39	.40				
12			.52	.47		
2			.51	.51		
8			.49	.56		
1			.45	.38		
6			.40	.46		
13					.55	.53
4					.50	.47
15					.48	.45
7					.43	.58
11					.39	.45

Note. S1 = Sample 1; S2 = Sample 2.

Model Testing: What Is It?

The purpose of structural equation modeling, as it relates to model testing, is to test the goodness of fit of one or more hypothesized models (to test the fit between the hypothesized relations between constructs [latent variables] and their observed variables that serve as indicators of those constructs).

What Tables Are Used?

In showing results of model testing, it is important to provide the reader with a visual representation of the model(s) being tested. Thus, figures frequently are used.

Two tables are commonly presented: (a) a table of intercorrelations among the variables included in the analysis and the means and standard deviations of those variables (see Tables 19.6 and 19.7) and (b) a table of the fit statistics (Table 19.8). In addition to these two types of tables, figures of the models often are included.

If only one model is being tested, a figure of the hypothesized model is often presented within the introduction section (see Figures 19.1 through 19.6). In addition, a figure of the structural model obtained from the results of the analysis is presented as well (see Figures 19.7, 19.8, and 19.9). If more than one model is being tested, then either all of the models are presented before (in the introduction) and after the analyses (in the Results section) or a figure of one hypothesized model and written descriptions of alternate models are presented within the introduction with only the best-fitting structural model presented within the Results section after the analyses.

If only one or two models are presented, then a table of fit indices is not required. This information can be incorporated within the text of the Results section.

"Play It Safe" Table

Figures 19.1 and 19.7 and Tables 19.6 and 19.8 are required to be comprehensive.

Example

Researchers wish to determine whether emotional expressive behavior (EEB), operationalized as crying, laughing, and yelling, is influenced by an individual's acceptance of EEB (attitudes and beliefs), which in turn are influenced by primary EEB influences (significant other's, father's, and mother's EEB) and personality (reactive and impulsive traits). They wish to test three models. Each of the three models is presented in a figure in this chapter. Models 2 and 3 are variants of Model 1 and do not need to be presented in the introduction, but they should be described within the text.

This is Model 1. This is a "Play It Safe" figure.

Figure 19.1

<u>Figure X</u>. Model 1 for emotional expressive behavior (EEB). Latent constructs are shown in ellipses, and observed variables are shown in rectangles.

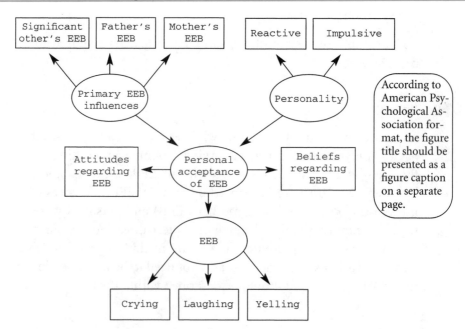

According to American Psychological Association format, the figure title should be presented as a figure caption on a separate page.

Figure 19.2

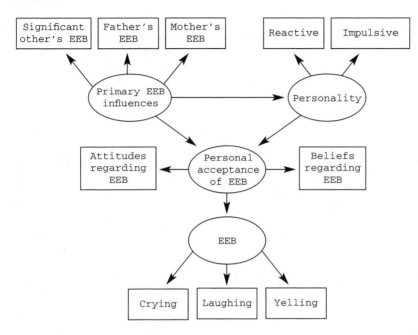

Figure X. Model 2 for emotional expressive behavior (EEB).

Arrows may cross, but try to avoid this. Figures should be as simple as possible to facilitate comprehension.

Figure 19.3

Figure X. Model 3 for emotional expressive behavior (EEB).

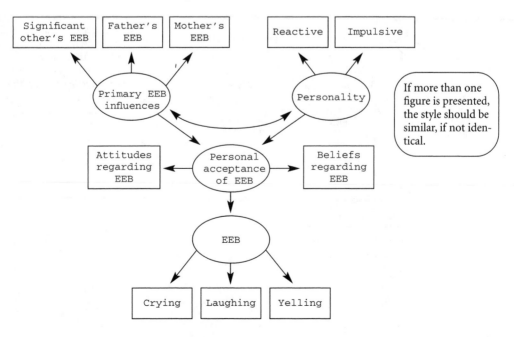

If more than one figure is presented, the style should be similar, if not identical.

Figures 19.4, 19.5, and 19.6 are variants of Model 1.

Figure 19.4

<u>Figure X</u>. Emotional expressive behavior (EEB), Model 1.

In this figure, the researcher wishes to focus attention solely on the latent variables.

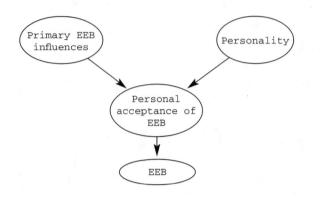

Figure 19.5

<u>Figure X</u>. Model 1 for emotional expressive behavior (EEB).

Measurement error terms are not shown.

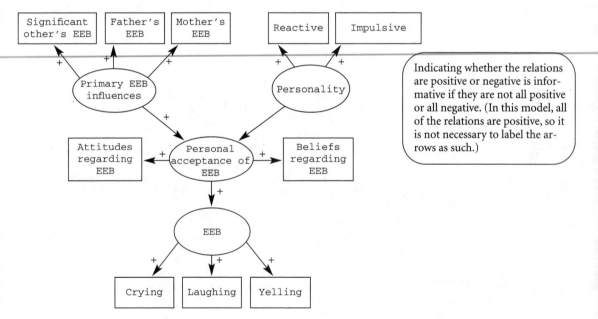

Indicating whether the relations are positive or negative is informative if they are not all positive or all negative. (In this model, all of the relations are positive, so it is not necessary to label the arrows as such.)

Figure 19.6

Figure X. Model 1 for emotional expressive behavior (EEB). I1 =
significant other's EEB; I2 = father's EEB; I3 = mother's EEB; P1
= reactive; P2 = impulsive; A1 = attitudes; A2 = beliefs; E1 =
crying; E2 = laughing; E3 = yelling. Latent constructs are shown
in ellipses, and observed variables are shown in rectangles.

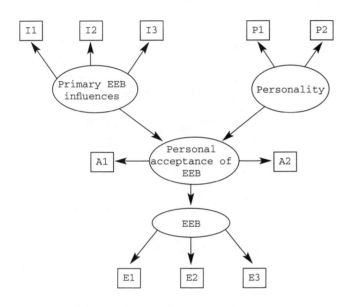

Using a symbol for the
indicator variable
rather than the name
of the variable can save
space if the diagram is
complex and contains
many latent or indica-
tor variables, or both.
The same symbols can
then be used in the
correlation table.

A table of intercorrelations among the variables along with the variable means and standard deviations is usually presented in the Results section.

Table 19.6

Table X

Descriptive Statistics and Zero-Order Correlations for Indicator Variables

Variable	1	2	3	4	5	6	7	8	9	10
Primary EEB influences										
1. Mother	--									
2. Father	26	--								
3. Significant other	.05	.09	--							
Personality										
4. Reactive	.67	.56	.44	--						
5. Impulsive	.34	.24	.11	.21	--					
Acceptance of EEB										
6. Attitudes	.65	.66	.23	.54	.21	--				
7. Beliefs	.06	.08	.15	.21	.12	.18	--			
EEB										
8. Crying	.76	.32	.21	.17	.19	.14	.28	--		
9. Laughing	.51	.45	.41	.43	.21	.43	.65	.34	--	
10. Yelling	.49	.37	.35	.34	.12	.48	.51	.29	.10	--
M	45.67	35.61	54.31	12.45	13.56	32.65	23.55	65.49	26.34	43.25
SD	8.45	7.86	9.54	1.34	2.54	6.48	4.55	10.12	2.36	8.31

Note. Correlations greater than .19 are significant at $p < .05$. EEB = emotional expressive behavior.

This table, along with a table of fit indices and illustrations of the models, is the "Play It Safe" table for model testing. This table provides headings to clearly identify the variables.

See chapters 7 and 13 for other examples of format for correlation tables and tables of means and standard deviations, respectively.

Labels in this table correspond with those in Figure 19.6.

Table 19.7

Table X

Zero-Order Correlations, Means, and Standard Deviations for Study Variables

Variable	I1	I2	I3	P1	P2	A1	A2	E1	E2	E3
I1	--									
I2	.09	--								
I3	.05	.26	--							
P1	.44	.56	.67	--						
P2	.11	.24	.34	.21	--					
A1	.23	.66	.65	.54	.21	--				
A2	.15	.08	.06	.21	.12	.18	--			
E1	.21	.32	.76	.17	.19	.14	.28	--		
E2	.41	.45	.51	.43	.21	.43	.65	.34	--	
E3	.35	.37	.49	.34	.12	.48	.51	.29	.10	--
M	54.31	35.61	45.67	12.45	13.56	32.65	23.55	65.49	26.34	43.25
SD	9.54	7.86	8.45	1.34	2.54	6.48	4.55	10.12	2.36	8.31

Note. Correlations greater than .19 are significant at $p < .05$. I1 = significant other's EEB; I2 = father's EEB; I3 = mother's EEB; P1 = reactive; P2 = impulsive; A1 = attitudes; A2 = beliefs; E1 = crying; E2 = laughing; E3 = yelling; EEB = emotional expressive behavior.

Table 19.8

Table X

Fit Statistics for Alternative Models

Model	df	χ^2	GFI	CFI	RMSR	IFI
1	44	187.65**	.81	.79	.78	.041
2	43	184.24**	.80	.80	.79	.044
3	43	79.23	.91	.91	.90	.021

Note. GFI = goodness-of-fit index; CFI = comparative fit index; RMSR = root-mean-square residual; IFI = incremental fit index.

**$\underline{p}$ < .01.

> Other indices of fit could be presented as well. It is best to present three or more indices.

> It is important to include a summary table of the fit indices, particularly if more than two models are being tested.

> This table, along with diagrams of the models and a table of correlations and means and standard deviations of the variables, is the "Play It Safe" table for model testing.

A diagram of the best-fit model could be included (in this instance, Model 3; see Figures 19.7, 19.8, and 19.9). Alternatively, three diagrams (one for each of the three models), including their standardized path coefficients, could be included.

Figure 19.7

Figure X. Standardized coefficients for Model 3. Latent constructs are shown in ellipses, and observed variables are shown in rectangles. EEB = emotional expressive behavior. *$\underline{p}$ < .05.

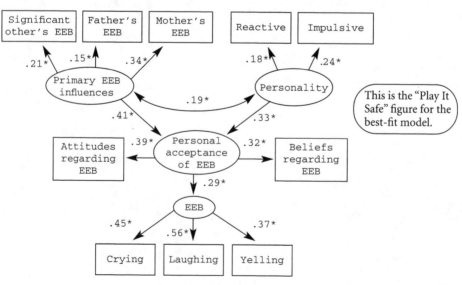

> This is the "Play It Safe" figure for the best-fit model.

Figure 19.8

Figure X. Standardized coefficients for Model 3 and their standard errors (in parentheses). All coefficients are significant at p < .05. χ² = 79.23; df = 43; p > .05; goodness-of-fit index = .91; comparative fit index = .91; root mean square residual = .90; incremental fit index = .021. Latent constructs are shown in ellipses, and observed variables are shown in rectangles. EEB = emotional expressive behavior.

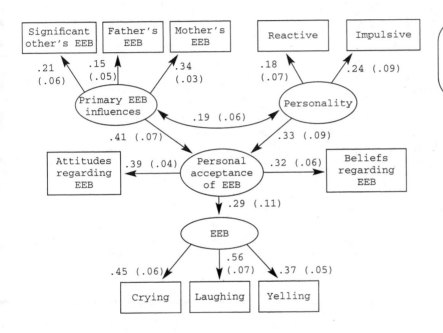

Figure 19.9

Figure X. Standardized coefficients for Model 3. GFI = goodness
of fit; CFI = comparative fit index; RMSR = root mean square
residual; IFI = incremental fit index. Latent constructs are
shown in ellipses, and observed variables are shown in rectangles. EEB = emotional expressive behavior. *p < .05.

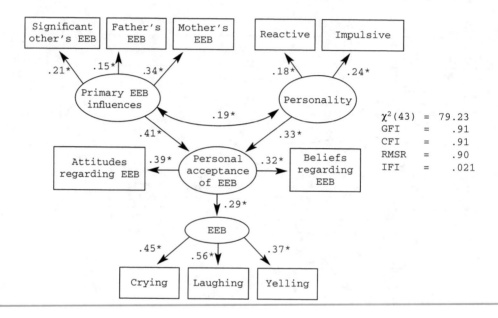

t Test of Means

What Is It?

A *t* test is used to determine whether a sample mean differs from a theoretical underlying distribution or whether two samples differ significantly from each other.

What Tables Are Used?

When there is only one *t* test to report, no table is required—the results are presented in the text rather than in a table. If results of several *t* tests are to be presented, then a table could be created. Tables 20.1 and 20.2 illustrate the results of multiple separate *t* tests.

"Play It Safe" Table

There is no comprehensive table for a single *t* test. Usually the means, standard deviations, *t* values, degrees of freedom, and significance level are reported within the text. However, a table can be presented for the results of several *t* tests. In this case, the most comprehensive table would include all of these statistics. Table 20.1

would be used if the degrees of freedom for each *t* test were the same, and Table 20.2 would be used if the degrees of freedom for each *t* test differed. Results for within-subjects *t* tests and between-subjects *t* tests are similarly presented.

Example

A researcher wishes to determine the effectiveness of a new mnemonic task on four different types of memory: memory for lists of faces, memory for lists of words, memory for lists of nonsense words, and memory for lists of two-digit numbers. The researcher conducts four experiments, one for each type of memory. Each study consists of a control group and an experimental group, with 10 participants in each group. The experimental group receives training on the mnemonic task for 2 hr, whereas the control group receives no memory training. For each study, the two groups of participants were matched on IQ, age, and grade point average. (Note that because the control and experimental groups were matched on IQ, age, and grade point average for each study, each of the four analyses is considered a matched-samples design.)

There is one independent variable, the mnemonic task. There are four dependent variables: memory for lists of faces, memory for lists of words, memory for lists of nonsense words, and memory for lists of two-digit numbers.

Exhibit 20.1

Independent Variable
1. Mnemonic task (program provided or no program provided)

Dependent Variables
1. Memory for lists of faces
2. Memory for lists of words
3. Memory for lists of nonsense words
4. Memory for lists of two-digit numbers

> This is the "Play It Safe" table for presenting the results of more than one *t* test.

Table 20.1

Table X

Group Differences for Memory Tasks Between Matched Groups Who Did
or Did Not Learn a New Mnemonic Task

Memory measure	No mnemonic task		Mnemonic task		
	M	SD	M	SD	t (9)
Faces	9.70	2.21	13.00	2.98	−4.33***
Words	7.60	1.51	10.00	1.70	−2.76*
Nonsense					
words	3.80	1.14	4.00	0.82	−0.51
Two-digit					
numbers	6.80	1.62	8.20	1.87	−1.74

*p < .05. ***p < .001.

Table 20.2

> This is the "Play It Safe" table for presenting the results of more than one *t* test when the number of participants is not the same for each *t* test.

Table X

Memory Differences Between Individuals Who Did Learn a Mnemonic
Task and Those Who Did Not Learn a New Mnemonic Task

Memory measure	No mnemonic task		Mnemonic task			
	M	SD	M	SD	df	t
Faces	9.70	2.21	13.00	2.98	9	−4.33***
Words	7.60	1.51	10.00	1.70	9	−2.76*
Nonsense						
words	3.80	1.14	4.00	0.82	9	−0.51
Two-digit						
numbers	6.80	1.62	8.20	1.87	9	−1.74

*p < .05. ***p < .001.

> Inclusion of the degrees of freedom as a column within the table is useful only if they differ for each analysis. In this example, however, Table 20.1 would be appropriate because the degrees of freedom are the same for all four variables.

CHAPTER 21

Word Tables

What Is It?

A word table provides descriptive or qualitative information. Definitions of variables, descriptions of referenced studies, and order of presentation of training phases are examples of the type of information included in word tables.

What Tables Are Used?

Usually, one table for each type of descriptive information is presented. Word tables are used sparingly; a word table should be included only if a thorough description is required and if a description in the text would be too confusing or too cumbersome.

"Play It Safe" Table

Because of the varied nature of word tables, there is no specific "Play It Safe" word table.

Example

Two researchers wish to determine which variables from a set are most strongly related to their Emotional Well-Being Scale in a population of individuals 65 years of age and older. There are a large number of variables, so the researchers determine that a word table is the most suitable way to present them.

Exhibit 21.1

Independent Variables
1. Spouse support
2. Children support
3. Friend support
4. Pet companionship
5. Routine living
6. Social activities
7. Physical activities
8. Mental activities
9. Income
10. Resource availability
11. Financial support

Dependent Variable
1. Emotional well-being

Table 21.1

Table X

Definitions of Variables and Sample Items

Variable	Definition	Sample item (positively keyed)
Spouse support	Spouse is emotionally supportive.	My partner is always willing to listen to me.
Children support	Child(ren) is/are emotionally supportive.	I can always count on my child(ren) to assist me with my problems.
Friend support	Friend(s) is/are emotionally supportive.	My friends call me on a weekly basis to keep in touch.
Pet companion-ship	Pet provides comfort.	I enjoy talking to my pet.
Routine living	Day-to-day living is predictable or repetitive.	I try to do something new every day.
Social activities	Individual participates in social activities.	I belong to a club where I can meet people.
Physical activities	Individual participates in physical activities.	I enjoy playing organized sports.

(Table X continues)

(Table X continued)

Variable	Definition	Sample item (positively keyed)
Mental activities	Individual participates in mental activities such as learning, reading, or problem solving.	I try to read every day.
Income	Level of income available to individual.	My average income per month is _____.
Resource availability	Number and quality of resources available to individual.	I have easy access to a grocery store.
Financial support	Financial support available from family or friends.	I can rely on family or friends when I have money problems.

Index

5750

About the Authors

ADELHEID A. M. NICOL is an assistant professor at the Military Psychology and Leadership Department at the Royal Military College of Canada. She obtained her BSc from McGill University and her MA and PhD from the University of Western Ontario. Her interests are in the areas of honesty–integrity testing, personality testing, test construction, legal issues in personnel selection, transformational leadership, and emotional intelligence. Dr. Nicol has taught courses in organizational behavior, research methods, social psychology, introductory psychology, and cross-cultural psychology.

PENNY M. PEXMAN is an associate professor of psychology at the University of Calgary. She received her PhD from the University of Western Ontario in 1998. In her research, she investigates the cognitive processes involved in word recognition, reading, and understanding figurative lanugage. Dr. Pexman has taught courses in introductory psychology, cognitive psychology, sensation and perception, and educational psychology and is the recipient of two teaching awards.